THE QUANTUM PHYSICS BIBLE

A FIREFLY BOOK

Published by Firefly Books Ltd. 2017

First printing

Publisher Cataloging-in-Publication Data (U.S.)

Library of Congress Cataloging-in-Publication Data is available

Library and Archives Canada Cataloguing in Publication

Clegg, Brian, author
         The quantum physics bible : the definitive guide to
200 years of subatomic science / Brian Clegg.
Includes index.
ISBN 978-1-77085-992-0 (softcover)
         1. Quantum theory--Popular works. I. Title.
QC174.123.C55 2017   530.12   C2017-902165-6

Published in the United States by   Published in Canada by
Firefly Books (U.S.) Inc.   Firefly Books Ltd.
P.O. Box 1338, Ellicott Station   50 Staples Avenue, Unit 1
Buffalo, New York 14205   Richmond Hill, Ontario L4B 0A7

Edited and designed by Whitefox

Printed in China

First published by Cassell,
a division of Octopus Publishing Group Ltd
Carmelite House
50 Victoria Embankment
London EC4Y 0DZ

Brian Clegg asserts the moral right to be
identified as the author of this work.

Editorial Director Trevor Davies
Production Controller Sarah Kulasek-Boyd

THE QUANTUM PHYSICS BIBLE

THE DEFINITIVE GUIDE TO 200 YEARS OF SUBATOMIC SCIENCE

BRIAN CLEGG

FIREFLY BOOKS

CONTENTS

INTRODUCTION

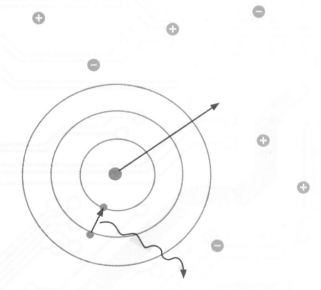

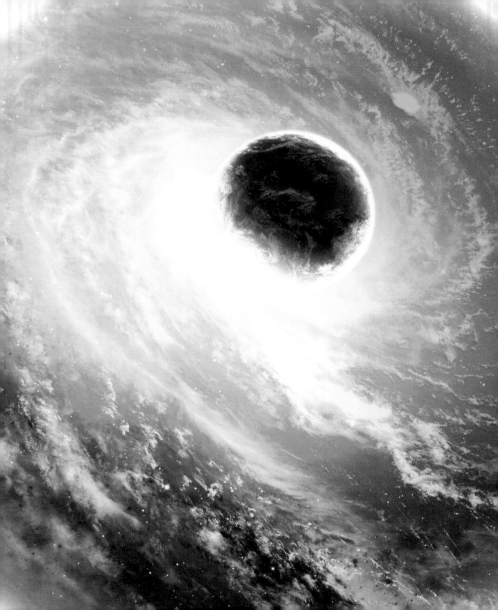

INTRODUCTION

Quantum physics intrigues because it has a mystery at its heart. This fundamental part of science describes the behavior of the atoms and subatomic particles that make up matter, as well as explaining the elusive but essential phenomenon of light. It lies behind electricity and magnetism — and most of the key inventions of the 20th and 21st centuries. Yet quantum physics describes a world in which events happen that are downright weird.

It's easy to think of particles such as electrons as tiny balls. Yet we know exactly what will happen to a ball if we throw it, given the details of its environment. The weirdness of the quantum world emerges from a lack of certainty. To the horror of some of the early developers of quantum theory, notably Albert Einstein, the equation describing the position of a particle after a certain amount of time deals only in probability. We can't state where it will be after 10 seconds, *just* the probability of finding it in various locations. And until we actually pin down the particle, *only* the probabilities exist. The same uncertainty is true for many other aspects of behavior at the quantum level.

Contrast this with a normal object over which probability apparently rules, such as a tossed coin. If we toss a coin, the chances are 50:50 of the outcome being heads or tails. However, once the coin is tossed, even if we haven't looked at it, just one side is facing up. It is either heads or tails. In the quantum equivalent, *only* those 50:50 probabilities exist until the particle interacts with the world around it. When the leading lights of quantum physics met at the Solvay Conference of 1927, this was still something that caused heated arguments among the participants. Some, such as Einstein and Erwin Schrödinger, were convinced that there had to be something underlying everything that was more "real," something not dependent on probability. Others, such as Niels Bohr and Werner Heisenberg, saw no need for such a reality. And it was Bohr and Heisenberg who have been proved right.

This may seem a matter of philosophical niceties. After all, the world around us and the things in it behave as we expect them to, despite being made up of quantum particles. So does it really matter what happens at the quantum level? The answer is a firm "Yes, it does!" It is this kind of weird quantum behavior that makes it possible for atoms to exist, makes it possible for the Sun to shine, and lies behind much of the technology we use today, from smartphones to lasers to MRI scanners.

Quantum physics has a reputation for being difficult — and the underlying mathematics is. But the basic concepts are easy to grasp, and essential if we really want to understand the quantum-based world around us. Before we take the plunge, though, it's important to know where this all came from — and to do that we need to look back at how our understanding of matter and light developed.

▲ The 1927 Solvay Conference brought together all the founders of quantum physics.

CHAPTER 1

THE UNCUTTABLE

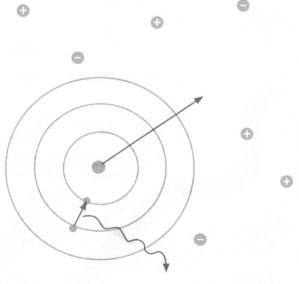

▶ The fresco **The School of Athens** *by Italian Renaissance artist Raphael brings together Greek philosophers in an imaginary gathering.*

THE NATURE OF EVERYTHING

Physics sits at the heart of science, underpinning our knowledge of the natural world. It's our way of describing and understanding how the essential stuff of the universe behaves. Which means that ever since human beings have thought about science, physics has played an important part.

This applies to every significant culture, although each began from a very different starting point. But because our modern scientific tradition is built on the foundations of Greek thought, it is most useful to look back to the Ancient Greek ideas that led to the development of modern physics and the transformation that came with quantum physics.

Much of the Ancient Greeks' science is no longer useful, but what they gave us was more valuable: a new way of looking at nature.

Thales
(ca. 624–528 BCE)

Pythagoras
(ca. 580–500 BCE)

Plato
(ca. 428–348 BCE)

Aristotle
(ca. 384–322 BCE)

THEOLOGY TO PHYSICS

• • • • • • • • • • • •

While everything natural had previously been considered the work of gods, the Greek philosopher Thales, born toward the end of the 7th century BC, and his successors looked for material rather than spiritual causes. Another Greek philosopher, Aristotle, writing more than 200 years later, called those who followed Thales *physikoi*, physicists, to distinguish them from theologians, the *theologoi*.

For the Ancient Greeks, physics meant the science of things on Earth, including living things, while astronomy was part of mathematics.

Euclid
(fl. 300 BCE)

Archimedes
(ca. 287–212 BCE)

Eratosthenes
(ca. 276–195 BCE)

Hipparchus
(ca. 190–120 BCE)

ELEMENTARY

The early Greek philosophers came up with the idea that matter was composed of elements. It was in the 5th century BCE that Empedocles first argued that everything was made up of earth, air, fire and water. This may seem overly simplistic, but it was devised on the scientific principle of basing a theory on observation. Empedocles had observed the way that, for example, when a piece of wood was burned it gave off air-like smoke, spouted fire, oozed water-like sap and ended up as earthlike ashes.

▲ Empedocles as imagined in a 17th-century engraving.

This theory was expanded by Aristotle, who added a fifth element, the quintessence, because he believed that everything above the Moon's orbit was perfect, and had to be made of something less worldly. The four elements were also used to derive an explanation of the way gravity worked.

Aristotle's approach to gravity was based on his idea that elements had a natural home. If something primarily consisted of earth or water, it wanted to be at the center of the universe, while fire and air moved away from the center of the universe. (Strictly, gravity was the tendency toward the center, while levity was the motion away.) This was one of the strongest arguments for the Sun orbiting the Earth, rather than vice versa.

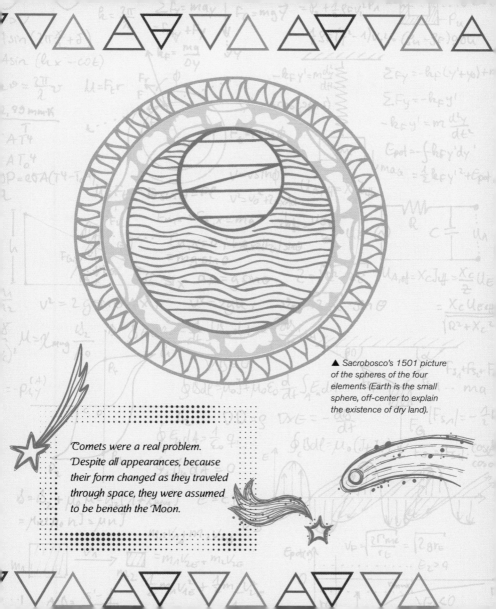

▲ Sacrobosco's 1501 picture of the spheres of the four elements (Earth is the small sphere, off-center to explain the existence of dry land).

Comets were a real problem. Despite all appearances, because their form changed as they traveled through space, they were assumed to be beneath the Moon.

ATOMISTS

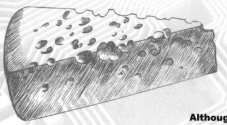

Although the four elements would remain the accepted theory of matter for around 2,000 years, it wasn't the only Ancient Greek approach.

Two of Empedocles' contemporaries, Leucippus and his pupil Democritus, came up with the surprisingly modern-sounding atomic theory. This was based on the idea that if you took anything — a piece of cheese, say — and cut it into smaller and smaller pieces, you would eventually get to the point at which it was impossible to cut further. They used a term for "not-cuttable": *a-tomos*. They had reduced substance to atoms.

These atoms were all made of the same basic stuff, but the atoms could have many different shapes, which defined the material. So cheese atoms, for example, would be different from air atoms.

In medieval times, an atom was also the smallest unit of time — 1/376th of a minute.

"Athomus, by þe diffinicion of trewe philisophres in þe sciens of astronomye, is þe leest partie of tyme. & it is so litil þat, for þe littilnes of it, it is undepartable & nei honde incomprehensible."

The Cloud of Unknowing, *14th century, anon*

ELEMENTS BEAT ATOMS

. . . .

Although atoms sound a more up-to-date approach than earth, air, fire and water, the four elements were arguably more scientific because they could be used to predict behavior. By contrast, because each substance had its own atoms, the atomic theory did not provide a better understanding of the world.

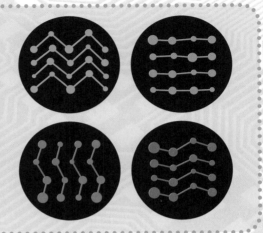

Aristotle rejected atomism, instead extending the four-element theory. As we have seen, he added a fifth element — the ether or quintessence — which helped challenge a fundamental principle of atomism. If everything were made of atoms, Aristotle argued, then there had to be a void between those atoms. This empty space or vacuum seemed an impossibility to Aristotle, as other material would naturally fill it; and if there were a vacuum, there was no reason why a moving object should ever come to a stop. In his elements model, there were no inconvenient voids.

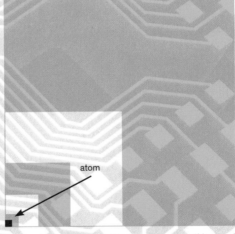

atom

▲ According to the Ancient Greek atomists, everything was made up of atoms that could not be cut.

THE MYSTERY OF LIGHT

Matter — physical stuff that you could touch — was relatively straightforward to describe, but ancient civilizations struggled more with the nature of light.

Originally it was only seen as the mechanism of sight, leading to some bizarre confusion where light was portrayed as a beam that emerged from the eye. But we now know that it is so much more, whether it's the means of transferring the energy from the Sun to the Earth that makes life possible, or the invisible web of photons that enables electromagnetism to work across a distance.

In prehistory, the brightest lights in the sky were awarded special importance. Many cultures considered the Sun and the Moon to be gods.

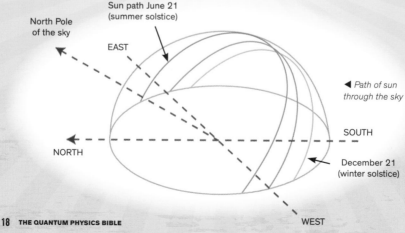

North Pole of the sky

Sun path June 21 (summer solstice)

EAST

◀ *Path of sun through the sky*

SOUTH

NORTH

December 21 (winter solstice)

WEST

ARCHITECTURE OF LIGHT

· · · · · · · · · · · · · ·

Though we can only guess at the precise use of Stonehenge, there is
no doubt that it, along with many other ancient structures, is aligned
with the Sun's rays at particular times of year. Although it is midsummer
that is most widely celebrated now at the site, there is evidence that the
key alignment occurs at midwinter, when the days start to grow longer
and the position of the light from the Sun tells of the coming spring and
renewal. Light was not only about sight, but a bringer of hope.

FIRE FROM THE EYE

Such was the enthusiasm among the Ancient Greeks to ensure that everything was made of the four elements that they assumed light must be a form of fire. Empedocles decided that this fire was emitted from the eye to reach the objects the viewer could see. He believed that the fire was carefully contained to avoid it

◀ According to Greek theory, vision required a stream of fire from the eye to reach the object that was seen.

coming into contact with the water in the eye and streamed out whenever the eye was open.

With a little logic, it seems obvious that this couldn't work, otherwise we could see in the dark (not to mention the lack of visible fire). But the Sun was still given a role as a mechanism to help the eye's fire to function. The Sun didn't provide the light that allows us to see, but activated its path. This strange theory would be modified over time, bringing in straight-line travel for light rays, but for centuries the eye's fire was still thought to have a role.

EASTERN LIGHT

It was only thanks to the influence of Arab philosophers around the start of the 11th century CE that fire from the eye was dismissed as unnecessary. The more rigorous Arab scientific ideas, including a better understanding of lenses and mirrors, were brought to the West over 100 years later.

WAVES AND CORPUSCLES

Over time, scientists realized that something must enable the light to travel from the Sun or a flame to a viewer's eye. By Isaac Newton's day in the later 1600s there were two opposing theories. Newton believed that light, rather like the atomists' matter, was made up of tiny particles that he called "corpuscles." Others, such as his Dutch contemporary Christiaan Huygens, preferred the idea that light was made up of waves, like the ripples that move over the surface of water.

Each theory had its problems. Using his corpuscles, Newton found it difficult to explain the way that light rays bent as they passed from air to glass, or why some light reflected off a piece of glass while some of it passed through. But this didn't mean that the wave supporters had it easy. They had to describe how waves could pass through empty space so that light from the Sun and the stars could reach us. That was easy for corpuscles, but waves had to have some material to wave in. There was a standoff between supporters of the two main theories for over 100 years.

▲ Like many early scientists, Huygens had wide interests. He studied light, astronomy, probability, mechanics… and still found time to invent the pendulum clock.

YOUNG'S SLITS

The argument between the wave and corpuscle theories was brought to a head by Thomas Young. Born in 1773, Young was a wealthy doctor who turned his mind to everything from the elasticity of materials to helping insurance companies decide on their premiums. But his greatest work was to demonstrate that light had to be a kind of wave, using an apparatus that became known as "Young's slits."

In 1801, Young described how he had produced two sharp beams of light from the same source and, using a pair of narrow slits, had caused the beams to overlap on a sheet of paper.

It was already known that waves could interfere with each other. If you drop two stones in water, at some points both waves will be moving up at the same time and will reinforce each other and make a stronger wave. At other points they will move in opposite directions and cancel each other out. Young saw fringes of dark and light on the paper — the light waves were undergoing this same interference process.

◀ Young was studying the effect of temperature on the formation of dewdrops, shining a candle light through a fine mist of droplets. The images of these droplets, projected on a white screen, formed colored rings around a white center. Young suspected the rings were caused by the waves of light interacting with each other, leading to his slits experiment.

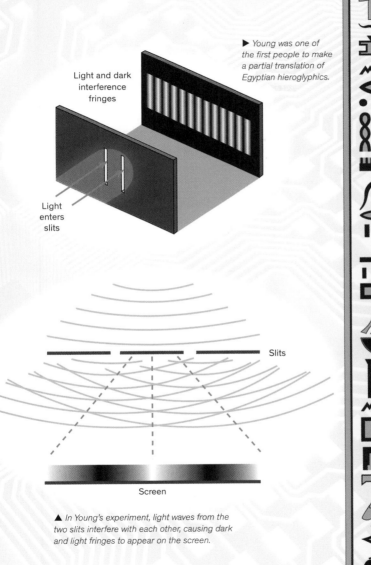

Light and dark interference fringes

Light enters slits

▶ Young was one of the first people to make a partial translation of Egyptian hieroglyphics.

Slits

Screen

▲ In Young's experiment, light waves from the two slits interfere with each other, causing dark and light fringes to appear on the screen.

ETHERIC WAVES

Although Thomas Young had clearly shown that light acted as a wave, his experiment did not explain how light traveled across the vacuum of space from the Sun.

By this time, it was known that sound waves could not cross space. When the air was sucked out of a jar containing a ringing bell, the sound could no longer be heard. But the bell could still be seen to move. The waves had to be vibrations in something else.

To explain how light managed this trick, scientists of the period modified the old idea of an invisible ether — something that filled all of space and could vibrate and form the waves of light. The ether was strange stuff — invisible, offering no resistance to anything passing through it, yet able to carry light across millions of miles from the Sun and stars.

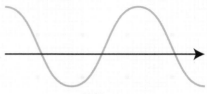

Light waves

Sea waves

▲ *Light waves move side to side compared with their direction of travel.*

When a wave passes through a material, the floppier the material, the more energy gets lost, so the wave dissipates more quickly. The ether had to be infinitely rigid... but still allow a wave through.

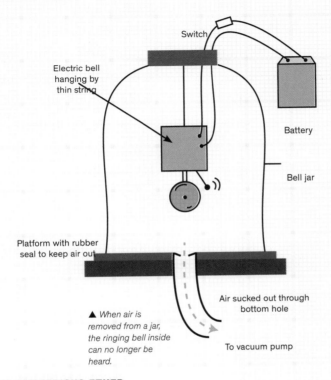

Switch

Electric bell
hanging by
thin string

Battery

Bell jar

Platform with rubber
seal to keep air out

Air sucked out through
bottom hole

To vacuum pump

▲ *When air is
removed from a jar,
the ringing bell inside
can no longer be
heard.*

THE MYSTERIOUS ETHER

The ether appeared to have a unique property. In the 19th century, features such as polarization (see page 142) showed that light was a side-to-side wave, like the ripples on the sea, as opposed to a compression wave, like sound. Side-to-side waves traveled only on the edge of a material — but somehow they passed through the middle of the ether. Despite this oddity, if light were a wave, it surely required something to wave through, and the ether made this possible.

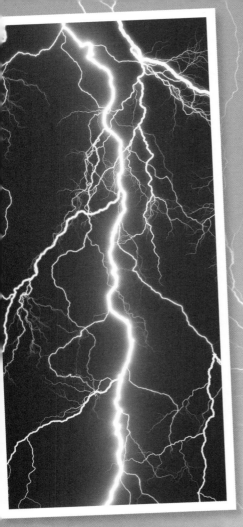

ELECTRICITY

Although matter and light make up much of the everyday world around us, there is another kind of stuff in the universe in the form of electricity. With its existence most obviously and dramatically demonstrated by a bolt of lightning, and by the unpleasant shock delivered by electric fish, there had long been an awareness of electricity.

Uncovering the exact nature of electricity would have to wait for the discovery of the component parts of the atom, but from the 1600s onward there was an increasing understanding of electricity as something distinct from the action of a magnet. Therefore it needed a distinct name, and the English natural philosopher William Gilbert coined "electricity" after the Greek word for

▲ *A typical lightning bolt carries as much energy as a medium-sized power station run for a second.*

The common term for a worker in science was "natural philosopher" until the late 19th century. "Scientist" was formed by analogy with "artist" in 1833, but took decades to become widely accepted.

amber, *electron*, via the later Latin "amber-like," *electricum*. Without it being clear exactly what was going on, electricity was increasingly used, first in demonstrations such as the "electric boy" who was suspended from silk ropes and used to carry an electric shock to a bystander, and later by scientists such as Michael Faraday, whose experiments led to the development of the electric motor.

AMBER'S ATTRACTION

The early Greek philosopher Thales noted the commonly observed phenomenon that when some materials, notably amber, were rubbed they attracted small objects to them, an early example of an electric effect.

Passing a current through an
"ELECTRIC BOY"
to cause a shock was a popular 18th-century experiment to entertain a crowd.

MAGNETISM

Like electricity, magnets were familiar from ancient times. Lodestones were naturally occurring objects capable of moving light pieces of some metals without touching them.

Because of their strange power, magnets were the subject of some of the oldest scientific studies. The earliest surviving document is Peter of Maricourt's 1269 work *Epistola de Magnete*.

In 1600, the same William Gilbert who named electricity published the highly influential book

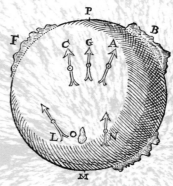

▲ *Gilbert used magnetized metal balls called "terrella" as models for the Earth's magnetism.*

▼ *Gilbert demonstrates magnetism to Queen Elizabeth, from* Hutchinson's Story of the Nations, *a 1920s history book.*

THE EARTH AS A MAGNET

Gilbert was the first to work through the practical implications of the Earth itself being a giant magnet. Not all of his ideas proved correct — he thought, for example, that gravity was the result of magnetism — but his approach was one of the earliest careful scientific examinations of a natural phenomenon. By using spherical magnets, Gilbert studied various oddities of compasses, from dipping needles to variations in the apparent position of the North Pole.

▲ The Earth's magnetic field, which extends far beyond the surface, enables compasses to align with the poles.

De Magnete. While crude magnetic compasses had been used for hundreds of years, the reason they worked was a mystery. The best suggestion at the time was that either the pole star Polaris or a strange island in Earth's far north somehow attracted magnets.

Gilbert is said to have spent the vast amount of £2,500 (US$3,250) on his experiments. In modern terms that's around £500,000 (US$650,000)

MESMER'S MAGIC FLUID

By the end of the 18th century, scientists were on the verge of a better understanding of electricity and magnetism, but many still associated these apparently magical phenomena with a kind of mysticism, making the timing ripe for the theories of German doctor Franz Mesmer.

Working in the 1770s, Mesmer first tried to cure a patient by getting her to drink a suspension of iron and trying to influence how it flowed through her body using magnets, but this initial approach was soon replaced by the more dramatic theory of animal magnetism.

Mesmer claimed that there was a natural magnetic fluid in the body and that it was possible for another person to manipulate this fluid (presumably as a result of their own magnetic powers). The manipulation of animal magnetism, often called Mesmerism, was supposed to cure the ill. Patients who were treated in this way reported feelings of heat in their bodies, went into trances and had fits.

It seems that Mesmer and his followers were producing the same kind of suggestive state as hypnotism — but, sadly for his patients, there was no magnetism involved... and no cure.

▲ German doctor Franz Mesmer devised the concept of animal magnetism.

Mesmer's ideas were investigated by a French Royal Commission including the unfortunate Joseph-Ignace Guillotin, whose name became attached to a killing machine he didn't invent. They decided Mesmerism was purely an effect of the imagination, but the report did little to stop the practice.

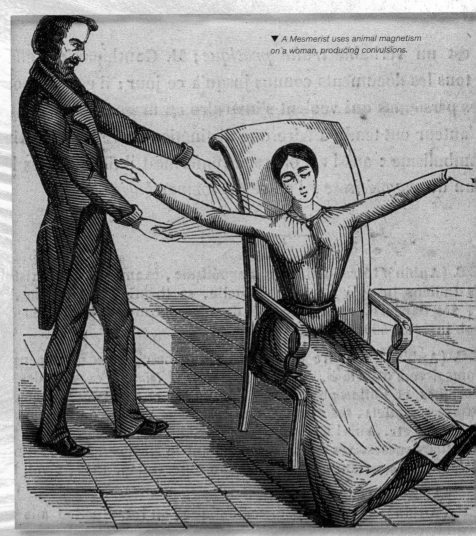

▼ A Mesmerist uses animal magnetism on a woman, producing convulsions.

MAXWELL'S SILVER HAMMER

$$\nabla \times E = -\frac{\partial}{\partial t} B$$
$$\nabla \times H = -\frac{\partial}{\partial t} D + J$$
$$\nabla \cdot D = \rho$$
$$\nabla \cdot B = 0$$

▲ *Maxwell's work resulted in four compact equations that detail the relationship between electricity and magnetism.*

The Victorian era saw the coining of a new term to replace "natural philosopher." Although "scientician," "scientman" and "savant" were all tried, it was "scientist" that took hold.

The initial force behind the change was probably the philosophers' distaste for something as practical as science, but it also reflected the development of the scientific professional, which saw figures such as Michael Faraday transform the understanding of electricity and magnetism.

The unification of the two phenomena as electromagnetism was triumphantly achieved by Scottish physicist James Clerk Maxwell. Faraday and his peers had shown how electricity could produce magnetism and magnetism electricity, but Maxwell developed a theory in which the two were tightly integrated.

▶ *The electromagnetic spectrum.*

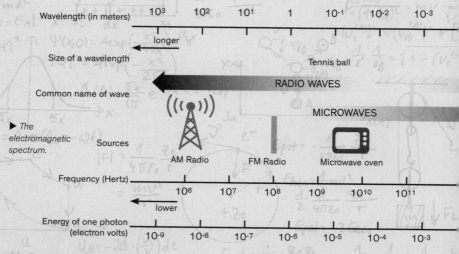

Wavelength (in meters)	10^3	10^2	10^1	1	10^{-1}	10^{-2}	10^{-3}

← longer

Size of a wavelength — Tennis ball

Common name of wave — RADIO WAVES — MICROWAVES

Sources — AM Radio — FM Radio — Microwave oven

Frequency (Hertz)	10^6	10^7	10^8	10^9	10^{10}	10^{11}

← lower

Energy of one photon (electron volts)	10^{-9}	10^{-8}	10^{-7}	10^{-6}	10^{-5}	10^{-4}	10^{-3}

He predicted that it should be possible to produce a self-sustaining wave of electromagnetism, but only at one particular speed: the speed of light. His mathematical analysis demonstrated that light could travel as an electromagnetic wave, and that such waves should also exist at different frequencies that were beyond the spectrum of visible light.

RADIO'S RECEPTION

Thirteen years after Maxwell published his breakthrough paper in 1864, the German physicist Heinrich Hertz produced electromagnetic waves — later known as radio waves — for the first time. Hertz's equipment was crude — an electrical spark jumping across a gap triggered waves in a pair of wires — but it was enough to produce a corresponding spark in a receiver across the room. Maxwell's theory and equations had made it possible to understand what light was made of.

▲ James Clerk Maxwell, with his wife Katherine, around the time of his work on electromagnetism.

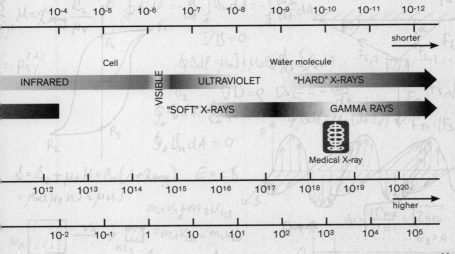

10^{-4} 10^{-5} 10^{-6} 10^{-7} 10^{-8} 10^{-9} 10^{-10} 10^{-11} 10^{-12}

shorter

Cell

Water molecule

INFRARED   VISIBLE   ULTRAVIOLET   "HARD" X-RAYS

"SOFT" X-RAYS   GAMMA RAYS

Medical X-ray

10^{12} 10^{13} 10^{14} 10^{15} 10^{16} 10^{17} 10^{18} 10^{19} 10^{20}

higher

10^{-2} 10^{-1} 1 10 10^{1} 10^{2} 10^{3} 10^{4} 10^{5}

THE END OF THE ETHER

For Maxwell, the idea that light was an electromagnetic wave did not do away with the need for the ether. He had envisaged his theory based on a mechanical model of the ether and could not let go of it. Others were bolder.

One nail in the coffin of the ether was an experiment originally designed to prove that it existed. Devised by American physicists Albert Michelson and Edward Morley, the experiment used the movement of the Earth. By 1887 Michelson and Morley had set a great slab of stone to float on a tray of mercury, so that the slab rotated smoothly at around a turn every six minutes. On top, a beam of light was set on paths at right angles. The movement of the Earth through the ether was supposed to produce a "wind" that would change the movement of the beam. As the slab rotated, it should have placed the two paths at different angles to the wind, making the light travel at different speeds down the paths, producing a shift of their respective waves when they eventually met. But nothing happened.

Descartes describes light as pressure in a "plenum" — a predecessor of the ether

Young's slit experiment seems to require an ether

1630   1679   1801   1817

Huygens describes light as waves in an ether

Fresnel suggests the Earth drags the ether with it to explain lack of astronomical variation

FIELDS BEAT THE ETHER

• • • • • • • • • • •

More daring physicists picked up on Faraday's idea of electrical and magnetic fields. Faraday had described a field as something that had varying values throughout space, similar to the contours on a map. They argued that light needed no ether, because it was a wave in the fields themselves.

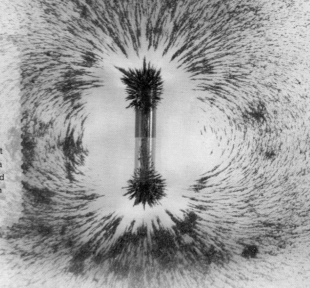

▲ *Faraday developed the concept of electrical and magnetic fields, demonstrated by the influence of a magnet on iron filings.*

Maxwell describes light as electromagnetic waves, but hangs onto the ether

Einstein's special relativity does away with the need for a fixed frame of reference like the ether

1864

1887

1905

Michelson–Morley experiment

CHAPTER 2

ENTER THE ATOM

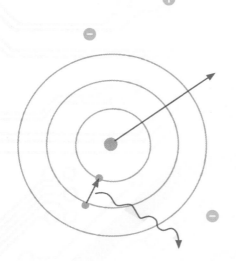

▶ The thin sheets of carbon in graphite reflect the atomic structure beneath.

DALTON'S ELEMENTS

When Leucippus and Democritus came up with the idea of atoms, these uncuttable components were of little scientific benefit. But around 1800 the British chemist John Dalton was responsible for a new take on atomic theory that could not just explain a lot that was observed, but also predict new possibilities for how different chemicals were combined and formed.

▲ An engraving of Dalton in his sixties.

Dalton brought back the Ancient Greek idea of atoms, but his version combined atomism with elements, requiring only a small number of building blocks rather than a different type of atom for every substance. Where Empedocles had used four elements, Dalton identified a wider range of natural substances, such as hydrogen, carbon and oxygen, as his building blocks for matter. These atoms combined in simple ratios to make more complex substances — such as two of hydrogen to one of oxygen combining to make water.

Although Dalton had no idea what an atom was (and many of his successors suspected it was no more than a handy model, rather than a real thing) he was able to discover the relative weights of a number of atoms, starting with the lightest, hydrogen, to which he allocated a weight of 1.

▲ Mural of John Dalton by Ford Maddox Brown at Manchester Town Hall, Manchester, UK.

Dalton developed his theory without a university education. As a Quaker he was barred from English universities, which accepted only members of the Church of England at the time.

ATOMIC INSPIRATION

Exactly how Dalton came to his idea is not clear. It might have been due to the way different substances combined in relatively simple ratios of weights, or to the physical behavior of gases when interacting with water.

▲ *In A New System of Chemical Philosophy, published in 1808, Dalton identified a range of atoms and simple molecules.*

THE ELECTRON

Although Dalton was insistent that atoms were, as envisaged by the Ancient Greeks, indivisible, something had been observed that seemed even lighter. A number of scientists had suggested that electrical charge was composed of the electrical equivalent of atoms, and by 1894 the Irish physicist George Stoney had come up with the name "electron" for the atom of electrical charge.

In parallel, other physicists had been working on cathode rays — the beams emitted by electrically charged plates in a near vacuum inside a glass tube. British physicist William Crookes discovered that these rays appeared to be made up of a negatively charged substance.

The definitive breakthrough was made in 1896, when British physicist Joseph John Thomson (universally known as J.J.) established that cathode rays were made up of a stream of individual particles, each with the same mass and charge. These particles seemed tiny, around 1/1,000th of the mass of a hydrogen atom. Exactly what these strange subatoms were was not clear, but the electron, which quickly became identified as the carrier of electrical charge, was here to stay.

▲ *J.J. Thomson became Cavendish Professor of Physics at Cambridge in 1884.*

◀ In this classic cathode-ray experiment, the rays are shone toward the glass at the near end of the tube, which glows green, except where the rays are blocked by the metal cross.

The "Crookes tube," with its beam of cathode rays steered by magnets, was the ancestor of all TVs and computer monitors until LCDs, LEDs and plasma flatscreens took over.

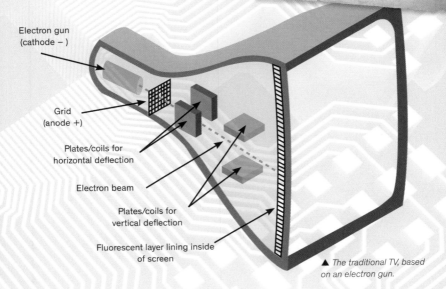

Electron gun
(cathode –)

Grid
(anode +)

Plates/coils for
horizontal deflection

Electron beam

Plates/coils for
vertical deflection

Fluorescent layer lining inside
of screen

▲ The traditional TV, based on an electron gun.

THE BROWNIAN ATOM

Remarkably, as late as 1905, many scientists thought that atoms didn't exist. They were happy with atomic theory, but thought this merely reflected the way elements interacted, without the need for the actual physical objects we call atoms.

Einstein wrote a series of brilliant papers in 1905, one of which would win him the Nobel Prize for Physics in 1921, but he did not get an academic post until 1908.

The work that persuaded many of the existence of atoms was not performed by a respected academic, but a clerk at the Swiss Patent Office by the name of Albert Einstein. Einstein had written a PhD thesis on the behavior of sugar molecules in solution that seemed to give backing to the idea that there were real tiny particles present, but before his thesis was submitted, he wrote a paper on Brownian motion.

Brownian motion describes the way that small particles, such as pollen grains, dance around when suspended in water. Einstein showed that this was due to individual water molecules colliding with the grains.

◀ *Although we usually see pictures of Einstein with white hair, he was only 26 in the key year of 1905.*

IT'S ALL IN THE NUMBERS

Einstein developed the mathematics to show that Brownian motion was as you would expect if water really were made up of tiny individual molecules. Although he hadn't strictly revealed anything about atoms, his work was taken as strong evidence that atomic theory was more than just a useful model.

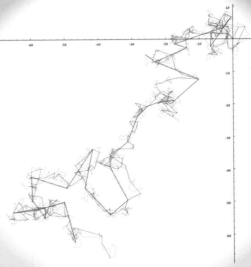

▶ A result of repeated collisions with water molecules, Brownian motion sees the visible particle make a series of apparently random moves.

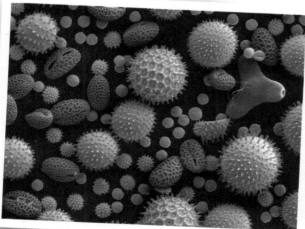

◀ Einstein explained the motion of suspended pollen, seen here under a scanning electron microscope.

THE PLUM PUDDING

A plum pudding is what we would now call a Christmas pudding. The "plums" were not plums at all, but raisins.

The Cambridge physicist J.J. Thomson, who established the existence of the electron, came up with a model for the atom that now seems very strange. He called it the plum pudding model, basing it on a popular dessert.

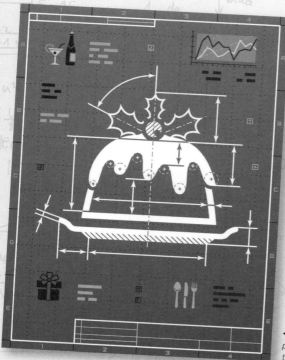

In this model the atom was made up of a "matrix" of positive charge, forming the pudding, with electrons as the "plums" scattered through the mix.

Perhaps because a sufficiently light positive equivalent of an electron hadn't been observed, Thomson assumed that the positive matrix was massless and that the electron plums made up the entire mass of the atom.

Thomson wrote in the *Philosophical Magazine* in 1904, "the atoms of the elements consist of a number

◄ *In the Christmas or plum pudding, raisins are distributed throughout the pudding mix.*

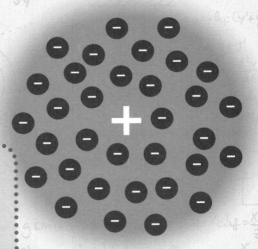

▶ In Thomson's model the negative electrons (he called them "corpuscles") were scattered through the positive matrix of the atom.

COUNTING PLUMS

As Thomson had already shown that the lightest atom, hydrogen, was at least 1,000 times heavier than an electron, he assumed that the hydrogen atom had to have at least 1,000 electrons in it. (With modern figures for the relative masses of an electron and hydrogen, there would be 1,837 electrons.) This is a considerable way from reality. In fact, a hydrogen atom contains only a single electron.

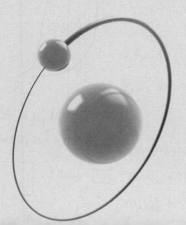

of negatively electrified corpuscles enclosed in a sphere of uniform positive electrification." Although his discovery, the electron, had been given that name 10 years before, Thomson always preferred to use the term "corpuscle" — a common name for particles since Newton's day. Exactly how the electrons behaved inside the pudding was not clear. Although they are usually drawn as if they are static, most attempts to describe them had them rotating in rings — even though this structure proved difficult to keep stable.

RUTHERFORD'S TISSUE PAPER

While Thomson developed his theory in Cambridge, Manchester-based New Zealand physicist Ernest Rutherford and his team were conducting experiments to explore the behavior of alpha particles. These were positively charged particles emitted by radioactive substances, too heavy to be the positive component of hydrogen.

Working with Hans Geiger and Ernest Marsden, Rutherford fired a stream of alpha particles at thin gold foil. The track of the particles was deduced by putting screens around the foil that lit up with a tiny flash when a particle hit them. An observer had to sit in the dark and peer at the screens to pinpoint the direction.

The team expected some alpha particles to be slightly deflected as they interacted with the diffuse positive charge of the "plum pudding" — and they were. But to everyone's amazement, other alpha particles reflected straight back off the foil. Rutherford realized that the gold atoms must have a small, concentrated positive charge to make the relatively heavy positive alpha particles bounce back.

BOUNCING SHELLS

Rutherford remarked that the reflection of the particles was like firing an artillery shell at a piece of tissue paper and having it bounce back.

▼ *Rutherford and Geiger (left) in the Manchester laboratory.*

NEW PHYSICAL
LABORATORY
THE OWENS COLLEGE
· MANCHESTER ·

▶ Architect's drawing of the Manchester laboratory.

Rutherford's laboratory in Manchester was state of the art. Purpose built in 1900, it had a ventilation system that drew outside air over oil baths to remove soot particles.

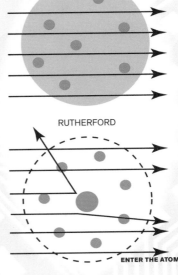

THOMSON

RUTHERFORD

▶ If Thomson's plum pudding model had been correct, alpha particles would pass straight through the atom, as the positive charge would be too diffuse to repel them, but some were reflected back.

THE MINIATURE SOLAR SYSTEM

There was something very appealing about the picture of the atom implied by Rutherford's discovery. At the atomic heart was the heavy, positively charged nucleus, while around the nucleus orbited the tiny electrons. It seemed that the atom was, in effect, a tiny solar system with the nucleus taking the role of the Sun and the electrons playing the planets.

This image, with its pleasing astronomical parallels, has come to dominate the visual representation of atomic structure, not least because it is simple and dramatic. Even official scientific bodies such as the International Atomic Energy Agency use

▲ Even the UN body the International Atomic Energy Agency uses the incorrect solar system model of the atom on its flag.

◄ It seemed very natural that the atom might have a structure that paralleled that of the solar system.

it. Unfortunately, no physicist could take it seriously as a model for the atom.

The trouble is that anything in orbit is accelerating (acceleration is a change of either speed or direction). And accelerating electrons lose energy in the form of light. So, if the solar system model were true, every atom would collapse as its electrons lost their energy and plummeted into the nucleus.

All kinds of structures were tried out to attempt to give the atom stability. For a while it seemed that it might be possible for the electrons to be fixed in place, kept stable by repelling each other, rather like an atom-scale crystal. This would mean there was no acceleration,

Rutherford borrowed the name "nucleus" from biology, where it was already used for the central "factory" of a complex cell.

but unfortunately a mechanical analysis of atoms with the observed numbers of electrons showed that they did not have a stable static configuration. Even without the problem of giving off energy, there seemed no way to stabilize moving electrons around an orbit. It would take the inspiration of a young Danish physicist, Niels Bohr (see page 78), to provide a solution.

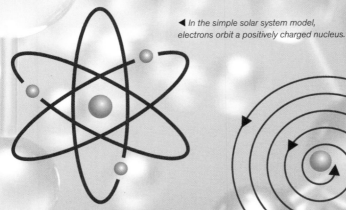

◀ *In the simple solar system model, electrons orbit a positively charged nucleus.*

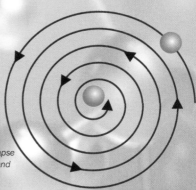

▶ *In reality, a solar system atom would collapse as the accelerating electrons give off light and spiral into the nucleus.*

ELEMENTS EXPLAINED

Although quantum physics would be needed to fix the solar system model, these new insights into atomic structure would be hugely valuable in a wider understanding of nature — and even to solve a mystery about the age of the Sun.

If we think of atoms as being like a special solar system, with the electrons like planets orbiting the nuclear sun, then we have a simple explanation for the behavior of the chemical elements. In his development of the periodic table, Russian scientist Dmitri Mendeleev had discovered that elements fitted into families that had similar chemical behavior. As atomic structure became better understood, scientists realized that atoms with similar numbers of electrons in their outer orbits shared behaviors.

▼ *In the periodic table, elements in the same column have similar properties and proved to have the same number of electrons in the outer orbit.*

H																	He
Li	Be											B	C	N	O	F	Ne
Na	Mg											Al	Si	P	S	Cl	Ar
K	Ca	Sc	Ti	V	Cr	Mn	Fe	Co	Ni	Cu	Zn	Ga	Ge	As	Se	Br	Kr
Rb	Sr	Y	Zr	Nb	Mo	Tc	Ru	Rh	Pd	Ag	Cd	In	Sn	Sb	Te	I	Xe
Cs	Ba		Hf	Ta	W	Re	Os	Ir	Pt	Au	Hg	Ti	Pb	Bi	Po	At	Rn
Fr	Ra		Rf	Db	Sg	Bh	Hs	Mt	Ds	Rg	Cn	Nh	Fl	Mc	Lv	Ts	Og

La	Ce	Pr	Nd	Pm	Sm	Eu	Gd	Tb	Dy	Ho	Er	Tm	Yb	Lu
Ac	Th	Pa	U	Np	Pu	Am	Cm	Bk	Cf	Es	Fm	Md	No	Lr

The idea of electrons in orbits (admittedly in a more complex form, where each orbit has a limit to the number of electrons it can hold) made it possible to explain how the different chemical elements varied.

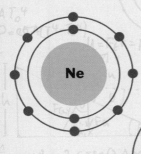

▲ Neon has the maximum eight electrons allowed in this orbit, making it less reactive. Carbon, with four electrons and four gaps, is a very flexible atom, enabling the complex structures required for organic forms.

Mendeleev predicted the existence of a number of elements that had not been discovered, because there were gaps in his table. He named them for the element above them — for example, ekasilicon, which was renamed germanium when discovered.

▲ Dmitri Mendeleev, who devised the periodic table of the elements.

THE ISOTOPE MYSTERY

When Dalton first worked out the relative masses of atoms, many had simple ratios, suggesting a structure of fundamental building blocks. Nitrogen, for example, was five times the weight of hydrogen. (The value is, in reality, seven times that of hydrogen. All Dalton's weights were initially incorrect due to the limitations of his equipment.)

His first list from 1803 had just five atomic elements — hydrogen, oxygen, azote (nitrogen), carbon and sulfur — though within five years he had 20. Later it was realized that the nucleus of an atom contained positively charged particles called "protons." As these accounted for most of the mass of an atom, this explained why atomic weight went up in relatively even units. But some elements had strange fractional ratios, such as chlorine, with an atomic weight 35.45 times that of hydrogen.

Another oddity provided an explanation. The English chemist Frederick Soddy, who studied radioactivity with Rutherford, had discovered that there were too many different atoms involved in a radioactive process to fit into the periodic table. He suggested that there could be variants of an element with different atomic

weights and radioactive behavior, but that were chemically identical. A friend of Soddy's, the Scottish doctor (and pseudonymous novelist) Margaret Todd, suggested calling these different versions of elements "isotopes," from a Greek term suggesting "equal place." But though different versions of the same atom had been identified, the reason for the differences in weight remained a mystery.

AN AVERAGE WEIGHT

The existence of isotopes explained the odd atomic weight of chlorine. The element contained different isotopes, mostly with atomic weight 35 and 37. These averaged out at 35.45.

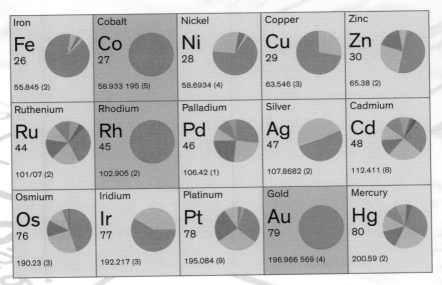

Iron **Fe** 26	Cobalt **Co** 27	Nickel **Ni** 28	Copper **Cu** 29	Zinc **Zn** 30
55.845 (2)	58.933 195 (5)	58.6934 (4)	63.546 (3)	65.38 (2)
Ruthenium **Ru** 44	Rhodium **Rh** 45	Palladium **Pd** 46	Silver **Ag** 47	Cadmium **Cd** 48
101/07 (2)	102.905 (2)	106.42 (1)	107.8682 (2)	112.411 (8)
Osmium **Os** 76	Iridium **Ir** 77	Platinum **Pt** 78	Gold **Au** 79	Mercury **Hg** 80
190.23 (3)	192.217 (3)	195.084 (9)	196.966 569 (4)	200.59 (2)

▲ *Examples of the relative occurrence of isotopes of sample elements.*

'Rutherford and Soddy used their knowledge of radioactive decay to develop the idea of dating objects by the state of their radioactive components, most commonly in radiocarbon dating.

◀ *A postage stamp printed in Sweden showing an image of Nobel Prize winner Frederick Soddy, ca. 1981. His 1921 Nobel Prize for Chemistry was awarded partly for his "investigations into the origins and nature of isotopes."*

NEUTRONS TO THE RESCUE

Although the existence of isotopes explained some oddities of atomic weight, they seemed to make a mess of the neat structure of the periodic table.

How was it possible to have different versions of the same element with varying masses, when an element needed to have a fixed number of protons in the nucleus to keep the right balance of protons and electrons, producing the expected chemical behavior?

The answer was to have neutral particles in the nucleus, making up the extra mass. Although the existence of these "neutrons" was first theorized by Rutherford in 1920, attempts to explain them as a linked proton and electron failed. It was only with the English physicist James Chadwick's discovery in 1932 of an uncharged particle with a similar mass to the proton that neutrons were understood to be stand-alone particles in their own right, occurring within the nucleus and explaining the varying atomic weights of isotopes.

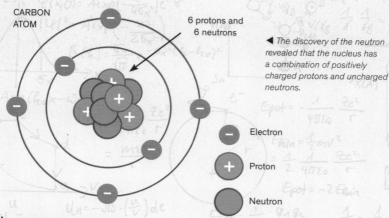

CARBON ATOM

6 protons and 6 neutrons

◀ The discovery of the neutron revealed that the nucleus has a combination of positively charged protons and uncharged neutrons.

Electron

Proton

Neutron

James Chadwick won the 1935 Nobel Prize for Physics for his discovery of the neutron.

Chadwick had been inspired by a new type of radiation observed in experiments in Germany and France. Like gamma rays (which proved to be high-energy photons), the new radiation did not respond to an electrical charge. Working at the Cavendish Laboratory in Cambridge, Chadwick aimed the new type of radiation at paraffin, which is a good source of protons. The radiation knocked protons out of the paraffin with energies that suggested that the radiation was made up of a stream of neutrally charged particles of similar mass to the proton.

Chadwick's neutrons were produced by bombarding elements such as lithium and boron with alpha particles.

SPLITTING THE ATOM

**When the Ancient Greeks devised the atom, splitting one was inconceivable —
atoms were what you got when you couldn't split matter any further. However, in
the early years of the 20th century Ernest Rutherford and Frederick Soddy had
shown that radioactivity, the spontaneous production of energy from a substance,
was a process in which some atoms naturally split apart.**

What had first been described as "alpha
rays," then renamed "alpha particles,"
turned out to be the positively charged
nucleus of a helium atom emitted from
the decaying nucleus of a radioactive
element. When, for example, radium
decayed to form lead, it went through a

serious of stages, producing different
elements as more and more chunks of the
original nucleus split off in a so-called
decay chain.

The discovery of the neutron led to
something much more dramatic. The
German physicist Otto Hahn, working
with Austrian physicist Lise Meitner, had
been firing neutrons at the heavy element
uranium in the hope that the neutrons
would "stick," producing an even heavier
atom. Instead, by 1938, just before the
Jewish Meitner left to escape the Nazis,
they discovered that they had split
uranium into a pair of lighter elements,
barium and krypton, in a process that
would become known as nuclear fission.
The following year, Meitner explained the
process theoretically.

◀ *The incoming neutron produces an unstable
isotope of uranium that splits into two lighter atoms
and three free neutrons.*

DDR

1879   1968

OTTO HAHN

$$^{235}_{92}U + ^{1}_{0}n \rightarrow ^{56}Ba$$

$$+ _{36}Kr + einige\ n$$

5

1979

▲ Otto Hahn won the 1944 Nobel Prize for Chemistry for his discovery of the fission of heavy nuclei.

▶ Lise Meitner was jointly nominated for the Nobel Prize with Otto Hahn, but was not awarded it. We can only speculate as to why this was the case, as no reasons are given by the Nobel organization.

◀ Fission was discovered with this apparatus, developed by Hahn with his assistant Fritz Strassmann.

CHAIN REACTIONS

Nuclear fission was impressive, but it was the debris of the reaction that gave it a new importance. A few years earlier, the Hungarian physicist Leo Szilard was thinking about a remark of Rutherford's that the energy released when breaking down an atom was "a very poor kind of thing" and anyone who thought it useful was "full of moonshine."

Szilard later said, "It suddenly occurred to me that if we could find an element which is split by neutrons, and which would emit two neutrons when it absorbed one neutron... we could sustain a nuclear chain reaction." Although no one knew of the possibility of nuclear fission, Szilard had proposed a way to harness it.

If more than one of the neutrons emitted when uranium decayed hit other uranium atoms, causing them to split too, then Rutherford's "poor kind of thing" would grow to an impressive level of energy. Hahn and Meitner soon realized that controlled nuclear fission could produce a steady and powerful flow of energy from a small amount of fuel. And if allowed to run away with itself, that energy could make a devastating bomb.

▲ Leo Szilard was joint author with Albert Einstein of a letter urging U.S. President Franklin D. Roosevelt to develop a bomb based on a nuclear chain reaction before Hitler could.

WAITING AT THE LIGHTS

Szilard said that the idea of a chain reaction came to him in a flash as he was standing in London's Russell Square, waiting for the traffic lights to change so he could cross the road.

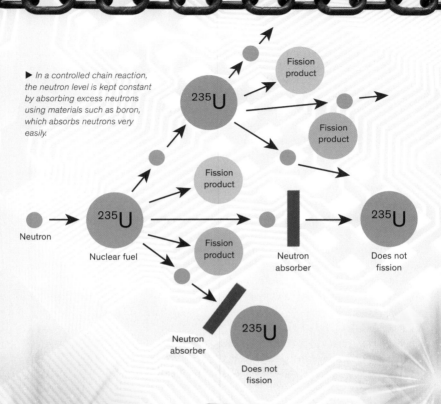

▶ In a controlled chain reaction, the neutron level is kept constant by absorbing excess neutrons using materials such as boron, which absorbs neutrons very easily.

235U

Fission product

Fission product

235U

Neutron

Nuclear fuel

Fission product

Fission product

Neutron absorber

235U

Does not fission

Neutron absorber

235U

Does not fission

In an attempt to impress Rutherford with his idea, Szilard described it using Rutherford's alpha particles rather than neutrons. Rutherford knew this wouldn't work (and was irritated that Szilard had patented his idea), so ignored him.

59

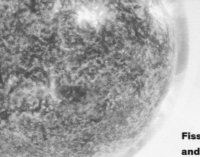

THE YOUNG SUN

Fission of the atomic nucleus may have provided a new and potentially terrifying source of power — but it had a competitor in the energy source of the Sun.

For centuries, there had been speculation about how the Sun could keep pouring out such vast quantities of energy. Both Darwin's evolutionary theory, which required long timescales for species to evolve, and geological evidence suggested that the Earth had existed for hundreds of millions of years. And the Earth without the Sun was unthinkable.

The Sun appeared to be a fire in the sky — but what was burning? The leading 19th-century physicist William Thomson, Lord Kelvin, suggested that the heat of the Sun was the result of the compression of the vast numbers of atoms in it being pulled together by gravity, much as a bicycle pump gets warm with the compression of air. But that would have given the Sun a lifetime of only 30 million years.

Kelvin's name would be permanently linked to temperatures when it was used for the scientific temperature scale, starting from absolute zero, with units of one kelvin equal to a degree Celsius.

1650

Ussher calculated the age of the Earth to be 6,000 years

1779

Comte de Buffon estimated the Earth's age from its cooling rate

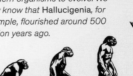

◀ *Darwin's* **Origin of Species** *made it clear that it must have taken many millions of years for modern organisms to evolve. We now know that* **Hallucigenia,** *for example, flourished around 500 million years ago.*

FOSSIL FUEL SUN

• • • • • • • • •

Earlier, Kelvin had calculated the lifetime of a fiery ball of coal the size of the Sun. Coal was the best fuel known at the time, but Kelvin found that it would burn out in just 20,000 years — far too quickly to be as old as the Earth.

◀ *A bicycle pump grows warm due to the heat generated by compressing the air.*

1854	1895	Current
Lord Kelvin calculated the Sun to be between 20 million and 100 million years old based on gravitational contraction	Perry recognized convection in the Earth's interior, thus dating the Sun to 1 billion years	Current understanding through radioactive decay is that the Earth is 4.5 billion years old

FANTASTIC FUSION

▲ *Nuclear fusion in stars produced the elements up to iron, but heavier elements are the result of vast stellar explosions called "supernovas."*

The answer to the long lifetime of the Sun proved to be nuclear fusion. In fission, the nucleus of an atom splits apart.

But it is also possible to join together two atomic nuclei to produce a heavier atom — as Hahn and Meitner had hoped to do (see page 56). This requires extremely high temperature and pressure, because atomic nuclei are positively charged and repel each other. However, if it were

possible to force the nuclei close enough, not only would they join together, or fuse, but something with far-reaching implications would occur.

Because of the complex interplay of the forces in the nucleus, when two nuclei fuse to produce a heavier one they can give off energy. Just as splitting a large nucleus produces energy, joining together small nuclei — most commonly fusing hydrogen nuclei to form helium — produces energy from nuclear fusion. With such a power source there finally was a mechanism to keep the Sun running for billions of years.

Fusion, conceived in the 1930s, depended on a new type of physics that had emerged at the start of the 20th century to explain a small discrepancy but soon transformed our understanding of the very small: quantum mechanics.

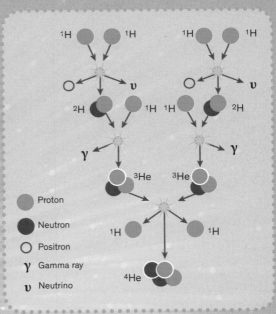

- Proton
- Neutron
- ○ Positron
- γ Gamma ray
- υ Neutrino

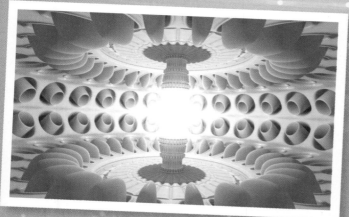

▲ In a star such as the Sun, hydrogen nuclei (protons) are fused together, initially producing a proton and a neutron and eventually merging to form a helium nucleus.

◄ Fusion reactors, such as the JET (Joint European Torus) at Culham in the UK, produce conditions on Earth for nuclear fusion to take place.

CHAPTER 3

CATASTROPHE AVERTED

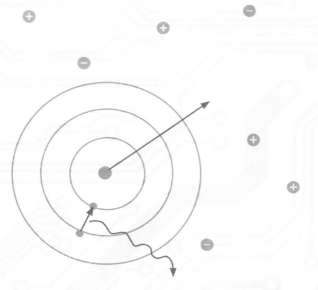

▶ The quantum age would be ushered in by
the inability of physics to explain what was
known as the "ultraviolet catastrophe."

A SMALL PROBLEM

In the previous chapter we saw how various pieces of the jigsaw puzzle of the nature of matter were slotted together. But by the time fusion was discovered, it had long been realized that there was something not quite right with "classical" physics, which assumed that everything at the level of atoms or electrons or light exhibited the same behavior as the objects and waves we see around us.

At the start of the 20th century one particular problem was causing confusion. It involved the way that hot objects glowed. Experimental findings ran so counter to what was expected that this would be given the (rather

color	approximate temperature		
	°F	°C	K
faint red	930	500	770
blood red	1075	580	855
dark cherry	1175	635	910
medium cherry	1275	690	965
cherry	1375	745	1020
bright cherry	1450	790	1060
salmon	1550	845	1115
dark orange	1630	890	1160
orange	1725	940	1215
lemon	1830	1000	1270
light yellow	1975	1080	1355
white	2200	1205	1480

▲ The color of heated metal provides a good measure of its temperature.

▼ As a piece of metal is heated, the light it emits gradually includes higher-frequency light waves.

overdramatic) name "the ultraviolet catastrophe."

As, say, a piece of iron is heated, it glows: first red, then yellow and white as more of the colors of the spectrum are added. The physical theory dealing with this predicted that as the frequency of the light waves increased, more energy should be given off. This would mean that everything, even at room temperature, should glow extremely brightly in the ultraviolet part of the light spectrum

A body that perfectly emits and absorbs electromagnetic radiation is known in physics as a "blackbody" even though it may be glowing brightly.

— that is, at higher frequencies than visible light. As this didn't happen, there clearly was something missing from the physical theory.

PIANO OR PHYSICS

Max Planck, the man who would take on the ultraviolet catastrophe, was born in Kiel, Germany, in 1858. At school he was accomplished at both science and music. When he came to choose a subject to study at university in 1874, Planck consulted the physics professor in Munich where his family now lived, Phillip von Jolly. The young student was uncertain whether to pursue music or physics.

Somewhat surprisingly, von Jolly recommended that Planck take the music option. According to von Jolly, physics was likely to prove a dead-end career, while musicians would always be needed. The professor explained that, with the exception of a couple of minor issues — one being the ultraviolet catastrophe — physics was pretty much complete and there was very little original work left to do.

To von Jolly's disappointment, Planck decided that he was perfectly happy to refine the details in physics and was not worried about coming up with anything original. In practice, though, Planck would light the fuse of a subject that would radically transform the physics of the very small.

◀ Although Planck gave up a career in music, he continued to be an excellent pianist throughout his life.

A stamp printed in Germany, ca. 1994, shows the discovery of quantum theory by Max Planck.

EUROPA

ENTDECKUNG DER QUANTENTHEORIE

$\Delta E = h \cdot \nu$

100

DEUTSCHE BUNDESPOST

Planck outlived all his children. His older son was killed in the First World War, both his daughters died in childbirth, and his younger son was executed by the Gestapo when he was implicated in a plot against Hitler.

▼ Despite von Jolly's warning, Max Planck went on to study at the Ludwig-Maximilians University in Munich.

PLANCK'S PACKETS

In 1900, Planck had a moment of inspiration that fixed the ultraviolet catastrophe. Existing theory incorrectly predicted that objects would pour out vast quantities of high-energy electromagnetic waves, but this depended on the assumption that light was like an ordinary wave that could be produced with any desired amount of energy. However, Planck found that if he pretended that light instead came in chunks — tiny packets of light called "quanta" — then the catastrophe went away, and the prediction of theory was an excellent match for what was ultimately observed.

For Planck's packets to deliver the correct result, they had to come with specific amounts of energy, dependent on the frequency of the light. This implied that there was a new constant of nature, now known as "Planck's constant" and represented by h. The energy in a packet of light was simply its wavelength multiplied by h.

Planck's constant proved intensely valuable in providing a link between the color of light (its wavelength or frequency) and the energy that light carries with it. It is an extremely small number — 6.626×10^{-34} — where 10^{-34} is 1 divided by 10,000 million trillion trillion. Max Planck would win the 1918 Nobel Prize for Physics "in recognition of the services he rendered to the advancement of Physics by his discovery of energy quanta."

The existence of quanta meant that nature had to be looked at in a different way. Instead of being entirely continuous with a possible value, some aspects, notably light, appeared to come in fixed-size chunks. This is similar to the way that coinage comes in fixed values, making it impossible to have 3.72 pennies.

▲ *Planck was already 42 when he solved the ultraviolet crisis, and he lived to the age of 89 .*

Traditionally, values in nature had been thought of as continuous, able to take any value. They are the equivalent of the position of a ball on a slope, where it can be at any level from top to bottom.

AN ACT OF DESPAIR

Max Planck was never comfortable with his solution to the ultraviolet catastrophe. He wrote: "The whole procedure was an act of despair because a theoretical interpretation had to be found at any price, no matter how high that may be."

In the quantum world, values in nature can only take certain values, making a step change from value to value. This is similar to the position of a ball on a series of steps, where it can only occupy certain specific heights, jumping from one position to the next.

THE PHOTOELECTRIC EFFECT

▲ A stamp, ca. 1965, printed in Sweden shows Philipp Lenard (left) and Adolf von Baeyer, both winners of Nobel Prizes in 1905.

The photoelectric effect is essentially the same as the photovoltaic effect used in solar cells, but the distinction is usually made that in the photoelectric effect the electron is pushed out of the material, while in the photovoltaic effect it stays in the material.

At around the same time as Planck was dramatically breaking light up into quanta, another German physicist, Philipp Lenard, made a strange discovery about the photoelectric effect.

This was a phenomenon first observed in the 1880s, when it had been discovered that shining light on certain metals seemed to blast out electrons. Early research used high-frequency ultraviolet light and established that the brighter this light was, the larger the amount of electricity produced.

Such a relationship fitted perfectly with Maxwell's theory that had established light as an electromagnetic wave (see page 33). It clearly predicted that the more intense a light beam was — whatever its color — the stronger the photoelectric effect should be, which was a natural consequence of light being a wave. But Lenard found that the color of the light did have an important influence on the outcome. In fact, if the light had too low a frequency — toward the red end of the spectrum or lower — it produced no effect at all, no matter how intense the light beam was. Once again, theory was struggling to match reality.

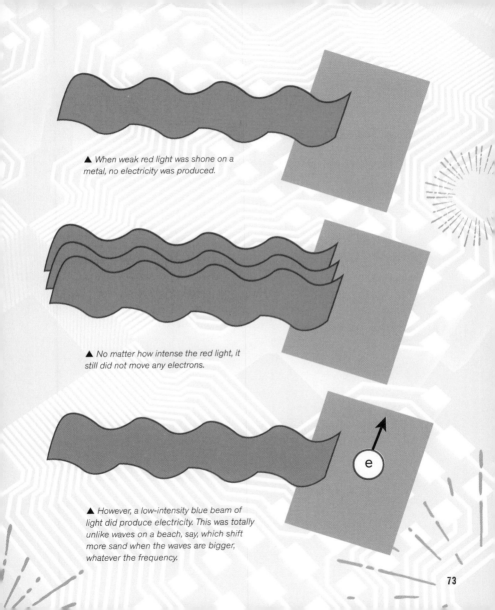

▲ When weak red light was shone on a metal, no electricity was produced.

▲ No matter how intense the red light, it still did not move any electrons.

e

▲ However, a low-intensity blue beam of light did produce electricity. This was totally unlike waves on a beach, say, which shift more sand when the waves are bigger, whatever the frequency.

EINSTEIN'S RADICAL PROPOSAL

The man who would explain Lenard's oddity and set the quantum physics revolution in motion was Albert Einstein. In another of the remarkable papers written in 1905, Einstein took Planck's imaginary packets and made them real.

Planck knew his packets of light were a trick to make the calculations work. Einstein dared to think that light *was* made of packets — particles that were later called "photons" — rather than waves. If this were the case, Lenard's observations suddenly made sense. Electrons were chunks of electricity, not variable amounts. If a single photon of light, rather than part of a continuous wave, had to knock an electron out of its place in the metal, the photoelectric effect would work only if that photon had enough energy. And Planck's work showed that light's energy was directly linked to the frequency — the color — of the light.

▲ Albert Einstein won the 1921 Nobel Prize for Physics for "his services to Theoretical Physics, and especially for his discovery of the law of the photoelectric effect."

EINSTEIN'S MISSED TARGET

Planck still wasn't happy to accept that quanta were real. When he proposed Einstein for the Prussian Academy of Sciences in 1913 he asked them to overlook the fact that Einstein "missed the target" with speculation such as that over light quanta.

BALMER'S SPECTRA

The idea that light came in chunks, rather than as a continuous wave, did not just deal with the ultraviolet catastrophe. It provided the key to the mystery of the work of Swiss mathematician Johann Jakob Balmer.

Since the 1860s it had been known that different chemical elements gave off specific colors of light when heated — so-called "spectral lines." Perhaps the most familiar everyday occurrence of this today is the strong yellow glow of a sodium-vapor streetlight. In 1885, when he was already 60, Balmer noticed that there was a pattern in the wavelength (or frequency) of light given off by hydrogen.

Although not a working physicist, Balmer managed to construct a formula that predicted the different frequencies of light that hydrogen would emit. It formed a regular pattern, though no one could

▲ Balmer worked all his life as a mathematician, but he is remembered only for a brief excursion into physics.

explain why. The answer would emerge from the next step forward in quantum theory, when a young Danish physicist had a remarkable idea about the structure of the atom.

▶ The element helium was discovered in the Sun from its spectral line by English astronomer Norman Lockyer, and named after the Greek word for the Sun, helios.

◀ The sodium vapor in streetlights produces the distinctive yellow frequencies associated with the element sodium.

BOHR FIXES THE ATOM

Niels Bohr was only 26 when he set off from Copenhagen for England, but he had already won the Danish Academy of Sciences' Gold Medal and published two papers in the *Philosophical Transactions* of the Royal Society. After a brief stay in Cambridge with J.J. Thomson, Bohr moved to Manchester to work with Ernest Rutherford. It was here that he began to think about the structure of the atom.

Rutherford had shown that there was a heavy positive nucleus with electrons outside it — but how did the electrons behave? The obvious solution was for the electrons to be like planets orbiting the nucleus (see page 48). But Bohr knew that the acceleration around the orbit would cause the charged electrons to spiral inward.

▲ *Niels Bohr came to the UK with a copy of* The Pickwick Papers *and a Danish-English dictionary to help him learn the language.*

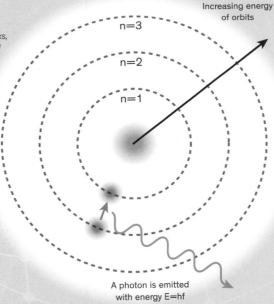

n=3

n=2

n=1

▶ *In Bohr's model, electrons could move only on fixed tracks, giving off light as they jumped between them.*

A photon is emitted with energy E=hf

He came up with the idea that electrons were effectively on fixed tracks around the atom and could not head inward. When the atom was heated, the electron could jump to the next available track, then fall back, giving off energy in the form of a photon of light. And when this new model was applied to hydrogen it fitted perfectly with Balmer's pattern.

▶ *A stamp printed in Denmark, ca. 1963, portrays Niels Bohr celebrating the 50th anniversary of his famous atomic theory.*

NIELS BOHRS ATOMTEORI
1913 - 1963

$h\nu = \mathcal{E}_2 - \mathcal{E}_1$

DANMARK   35

V. BANG del.   CZ SLANIA sc.

THE QUANTUM LEAP

Bohr's atom was a quantum object. The electron could not have any amount of energy, but was limited to fixed orbits. Each jump from orbit to orbit was called a "quantum leap."

Bohr's original mathematics worked only for the hydrogen atom, but he had set the direction for a better understanding of the workings of the tiny particles that make up reality. However, there was a price attached. The quantum leap had no equivalent in the world we see around us. The electron did not move gradually from one orbit to another, as a spacecraft does as it travels around the Earth.

Instead it jumped instantaneously from one orbit to the other with no gradual change.

Quantum physics was beginning to show its dark side. The theory worked and explained what it had not been possible to explain before. But it required a different kind of leap — a leap of imagination.

During the Second World War, nuclear scientists were ordered to travel under assumed names to minimize their security risk. Niels Bohr was assigned Nicholas Baker, known to his colleagues as Uncle Nick. Bohr's handwriting was so bad that there was wide dispute over whether a letter from him was signed Niels Bohr or Uncle Nick.

A TINY LEAP

• • • • •

Although in popular usage "quantum leap" has come to mean a big change, in physics it is the smallest possible difference.

▶ *When a spacecraft changes orbit, it moves gradually from one orbit to another.*

PARTICLE AND WAVE

Einstein and Bohr had both made huge leaps forward in explaining physical phenomena by assuming that light was made up of particles — photons — just as Isaac Newton had once suggested in his corpuscle model (see page 21). But in fixing one set of problems, they had opened up a whole new can of worms. After all, there was plenty of experimental evidence that light behaved as if it were a wave. And Maxwell's electromagnetic theory *required* light to be a wave.

Which was it, wave or particle? Bohr had already taken the step of allowing electrons to behave like no object with which we are familiar, able to make quantum leaps from one orbit to another. Now he and his colleagues took another step away from the everyday world. Light, they decided, was *both* wave and particle, yet also *neither*. At any one time it could behave as if

it were a particle or a wave, but never both at the same time. If an experiment required it to be a wave, it acted like a wave — and the same for particles. The principle would become known as "wave–particle duality."

Although it may seem little more than a pragmatic compromise, this was a significant step in the transformation of the understanding of science.

◀ *Arthur Eddington was an expert on the general theory of relativity and a leading science popularizer of his day.*

Historically, science was seen as an attempt to uncover the truth — the absolute factual essence of aspects of nature (and it is still often seen this way by the public). However, modern science is primarily concerned with developing models — functional descriptions, often mathematical, that produce results paralleling those observed in nature. It had previously been assumed that light *was* a wave, or later that it *was* a particle. It would become clear that both of these descriptions were models.

Describing this duality in 1928, English physicist Arthur Eddington wrote, "perhaps as a compromise, we had better call it a 'wavicle.'" Thankfully, despite making it into the Oxford English Dictionary, the word hasn't caught on.

THE DOUBLE SLIT REVISITED

The biggest challenge to thinking of light as particles is Young's slits (see page 22). Here, a pair of waves interfere with each other, producing a pattern that shows that some waves are reinforcing each other, producing bright fringes, and elsewhere canceling each other out, producing dark fringes.

It's hard to see how this could possibly happen if light acts as a particle. Yet, remarkably, it would be experimentally demonstrated that it *is* possible. As technology became more sophisticated, it became possible to send light toward

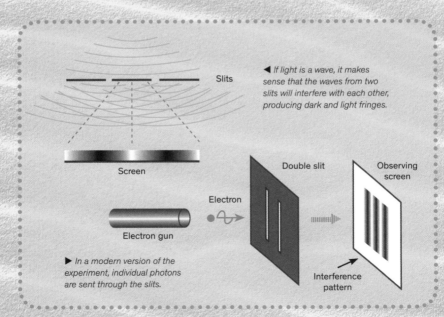

◀ *If light is a wave, it makes sense that the waves from two slits will interfere with each other, producing dark and light fringes.*

Slits

Screen

Double slit

Observing screen

Electron

Electron gun

▶ *In a modern version of the experiment, individual photons are sent through the slits.*

Interference pattern

the slits a single photon at a time. When the photon reached the screen, it produced a tiny bright dot. By keeping the dots lit, it was possible to build up a picture over time, photon by photon. The result was a dotty version of the same pattern of bright and dark fringes. Somehow, the quantum particles produced a wavelike interference effect as they passed through the equipment.

Although the idea of a detector being used to check which slit the photon passed through was proposed early in the development of quantum theory, it couldn't be tried out until the 1970s, because it required a detector that did not absorb the photon.

LOOK AND IT GOES AWAY

To make matters even more puzzling, the interference pattern does not form if a detector is used to check which of the two slits the photons pass through.

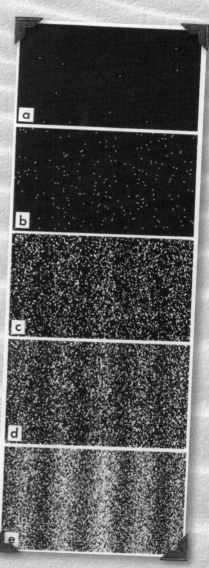

▶ The result, as *individual photons pass through, is the buildup of an interference pattern, dot by dot.*

83

MATTER WAVES

A French physicist, the splendidly titled Louis Victor Pierre Raymond, duc de Broglie, completed the wave–particle duality picture by reversing it. If light, which had been thought of as a wave, could behave like a stream of particles, perhaps particles, such as electrons, could behave as if they were waves.

De Broglie suggested this in 1924, and it was demonstrated just three years later by the American duo of Clinton Davisson and Lester Germer. They made use of an effect that occurs whenever a wave hits an obstacle — the wave bends around the edge of the obstacle. This is why we can hear someone around the corner of a building. If you fire a beam of particles at an obstacle, they will reflect off it at various angles. But a wave will tend to pass around the obstacle, bending back inward

▲ De Broglie inverted the idea of photons to conceive of particles behaving like waves. He won the 1929 Nobel Prize for Physics for his discovery.

▶ U.S. physicists Davisson and Germer with their electron-diffraction equipment.

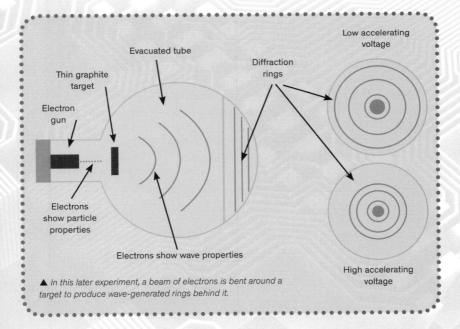

Evacuated tube

Thin graphite
target

Electron
gun

Electrons
show particle
properties

Electrons show wave properties

Diffraction
rings

Low accelerating
voltage

High accelerating
voltage

▲ *In this later experiment, a beam of electrons is bent around a target to produce wave-generated rings behind it.*

on the other side. In their experiment, Davisson and Germer found that the path of a stream of electrons was bent inward by a crystal in the same way. Later experiments reproduced Young's slit experiment using electrons rather than photons — and, once again, they behaved as waves.

De Broglie's concept illustrated just how far quantum physics was beginning to undermine old assumptions. Electrons were constituents of matter — of stuff — while

photons were insubstantial light. Yet they behaved in some circumstances as if they were waves and in others as particles. As the quantum world emerged, the old distinctions were becoming less certain.

Louis was the seventh duc de Broglie, a title now held by Philippe-Maurice, the ninth Duc. The family was originally Italian (and called Broglie), immigrating to France in the 1640s.

ATOMIC WAVES

The idea that electrons could behave like waves fitted well with Bohr's model of the atom. Here, an electron could occupy only specific orbits around the nucleus, as if it were running on tracks — and jumped between these orbits in quantum leaps as it gained or lost energy in the form of a photon. But what determined which orbits were acceptable and which were not?

If an electron could act like a wave, it was possible to imagine the wave of the electron running around the orbit. Most orbits would not have a circumference of an exact number of wavelengths, meaning that the wave could not return to a notional starting point. But if the wavelength were an exact fraction of the circumference, then the wave would fit the orbit exactly. It seemed that wave–particle duality was strongly linked to the structure of the atom.

The wave pattern of an electron around an atom is known as an "orbital."

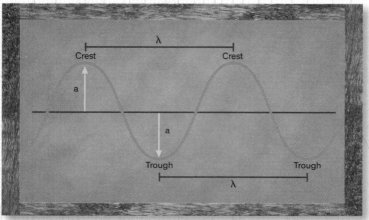

◀ The wavelength (λ) is the distance from a point on the wave to the equivalent point on the next peak or trough. The distance "a" is the size of the wave, known as its amplitude.

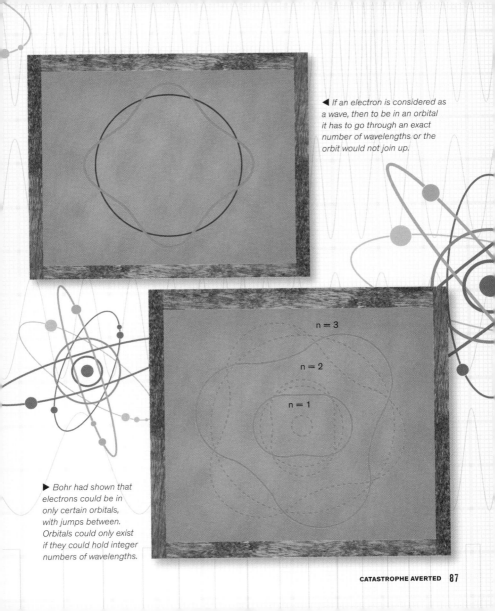

◀ If an electron is considered as a wave, then to be in an orbital it has to go through an exact number of wavelengths or the orbit would not join up.

n = 3

n = 2

n = 1

▶ Bohr had shown that electrons could be in only certain orbitals, with jumps between. Orbitals could only exist if they could hold integer numbers of wavelengths.

CHAPTER 4
QUANTUM REALITY

▶ Laser light display.

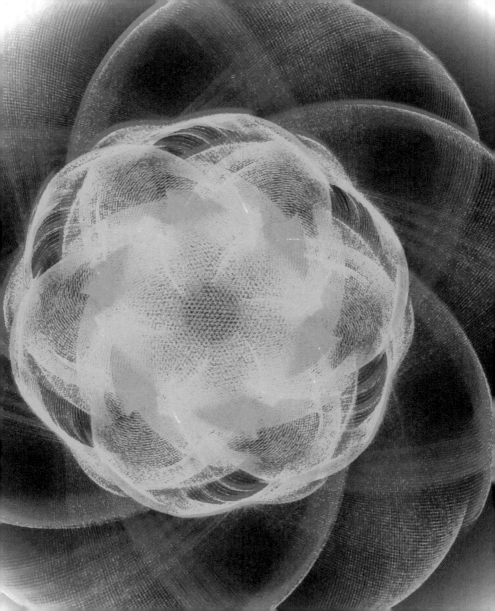

MATRIX MECHANICS

Although Niels Bohr had managed to construct an effective model of the hydrogen atom using quantum concepts, it proved difficult to extend to a wider picture of the whole world of matter and light.

A young German physicist, Werner Heisenberg, felt that Bohr's picture was too tied to the conventional world of everyday things — it was based on a model of how planets worked, with one fundamental tweak in the form of the quantum leap. Heisenberg believed that it would be better totally to detach his thinking from "visible world" parallels.

Instead, Heisenberg developed a mathematical model of quantum particles known as "matrix mechanics." Heisenberg's model did not provide any analogy to help our understanding of the quantum world — it was nothing more than a collection of numbers. Yet these numbers worked very effectively in predicting the behavior of quantum particles.

ENTERING THE MATRIX

Heisenberg's method involved manipulating matrices: two-dimensional arrays of numbers that behaved unlike familiar numbers. With matrices, for example, A × B did not equal B × A. This branch of mathematics was unfamiliar to most physicists, who treated it with suspicion.

▲ Werner Heisenberg won the Nobel Prize for Physics in 1932 for "the creation of quantum mechanics."

Unlike most quantum physicists, Heisenberg stayed in Germany to work for the Nazis in their attempt to construct a nuclear weapon.

▶ A matrix is a two-dimensional array (and the math can seem confusing). The top-left result comes from multiplying the top row of the first matrix by the left column of the second matrix and adding them together — in the example highlighted $(1 \times 5) + (0 \times -5) = 5$.

$$\text{For} \quad A = \begin{bmatrix} 1 & 0 \\ 2 & 1 \end{bmatrix} \quad \text{and} \quad B = \begin{bmatrix} 5 & 4 \\ -5 & 1 \end{bmatrix}$$

$$AB = \begin{bmatrix} 1 & 0 \\ 2 & 1 \end{bmatrix} \begin{bmatrix} 5 & 4 \\ -5 & 1 \end{bmatrix} = \begin{bmatrix} 5 & 4 \\ 5 & 9 \end{bmatrix}$$

$$BA = \begin{bmatrix} 5 & 4 \\ -5 & 1 \end{bmatrix} \begin{bmatrix} 1 & 0 \\ 2 & 1 \end{bmatrix} = \begin{bmatrix} 13 & 4 \\ -3 & 1 \end{bmatrix}$$

SCHRÖDINGER'S EQUATION

Heisenberg was not the only young physicist to be fascinated by the quantum revolution unleashed by Planck, Einstein and Bohr. Austrian scientist Erwin Schrödinger was also battling to extend quantum physics to cover all particles. However, unlike Heisenberg, Schrödinger felt that it had to be possible to visualize a model for it to be useful.

Rather than start from a purely mathematical viewpoint, Schrödinger followed Bohr and de Broglie in stressing the importance of regarding quantum particles as waves, and he developed a parallel theory to Heisenberg's called "wave mechanics." In this approach, the quantum particle was treated as a wave, and Schrödinger developed an equation that he hoped would predict the behavior of a quantum particle over time.

For the moment there seemed to be two fundamentally different approaches, but Schrödinger's equation, impressive though it was in theory, had two big

$$ih\frac{\partial}{\partial t}\,\Psi(\mathrm{r},\,t) = \hat{H}\Psi(\mathrm{r},\,t)$$

▲ *In this representation of Schrödinger's equation, both Ψ (the Greek letter psi) and Ĥ (pronounced "H hat") are compact representations of mathematical structures.*

▶ *Erwin Schrödinger shared the 1933 Nobel Prize for Physics with Paul Dirac for "the discovery of new productive forms of atomic theory."*

problems. First, the equation seemed to say that over time particles would occupy a larger and larger space, spreading out as if they were inflating. This was not observed, thankfully, as matter would tend to fall apart as a result. Yet it seemed a fundamental part of the way that the equation's predictions developed over time. And second, the equation contained the value "i" — and this appeared to deliver an imaginary result.

RESOLVING THE CONFLICT

The English physicist Paul Dirac would later show that Heisenberg's and Schrödinger's approaches were different representations of the same thing and exactly equivalent.

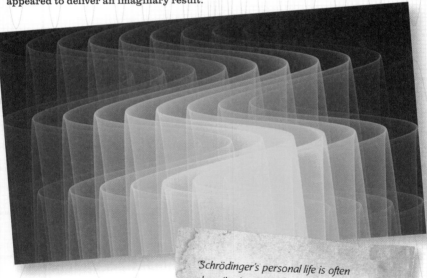

Schrödinger's personal life is often described as "unconventional." When escaping Nazi Germany for Ireland in 1938, for example, he was accompanied by both his wife and a Mrs. Hilde March, who had had a daughter by Schrödinger in 1934.

IMAGINARY PHYSICS

Schrödinger's equation produced an imaginary result in the mathematical sense. Centuries before, mathematicians had pondered what value the square root of −1 could have. Multiplied by themselves, both positive and negative numbers produce positive numbers: $1 \times 1 = 1$ and $-1 \times -1 = 1$. So what number when multiplied by itself would produce a negative number? The square root of −1 was arbitrarily given the representation i. From the basic idea that $i \times i = -1$, mathematicians extended the numerical horizon.

The phrase "imaginary numbers" was first used as a term of derision by the 17th-century French philosopher René Descartes.

MAKING IT REAL

As long as the final outcome of a calculation was real, you could work with imaginary numbers to model two-dimensional behavior, such as that of a wave. But the problem with Schrödinger's wave equation is that the final result features an imaginary number — and no one could conceive what that meant.

In conventional arithmetic, -1 can be produced only by multiplying two different values. What could be multiplied by itself to produce -1?

$1 \times -1 = -1$	$-1 \times -1 = 1$
$1 \times 1 = 1$	$-1 \times 1 = -1$
$? \times ? = -1$	

Initially this was mathematical play — but this apparently useless piece of math was shown to have a practical application. If real numbers represented the horizontal axis of a chart and imaginary numbers the vertical axis,

$$i\hbar \frac{\partial}{\partial t} \Psi(\mathrm{r}, t) = \hat{H}\Psi(\mathrm{r}, t)$$

◀ The innocent-looking "i" at the start of the equation seems to bring the imaginary into the real world.

Although the Ancient Greek mathematician Hero of Alexandria did appear to use an imaginary number accidentally, they were first used deliberately by Italian mathematicians in the 16th century as they studied the solutions of equations involving cubes and fourth powers.

points in two-dimensional space could be labeled with a single "complex" number such as $3 + 2i$. And these numbers proved extremely useful in physics and engineering.

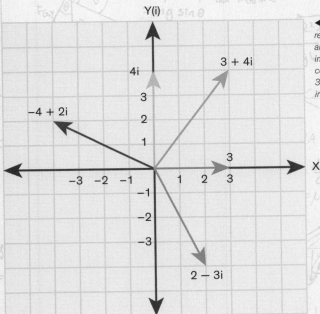

◀ With the horizontal axis representing real numbers and the vertical axis imaginary numbers, a single complex number such as $3 + 4i$ represents a position in two-dimensional space.

BORN'S PROBABILISTIC SOLUTION

The imaginary-number problem was solved relatively quickly when it was realized that it was the *square* of the value produced that made the equation work properly, which disposed of the imaginary number, as the square of an imaginary number is, by definition, a real negative number. But this fix still didn't deal with the way that the equation seemed to predict that a quantum particle would spread out, getting bigger and bigger over time — something that clearly didn't happen.

This expanding-particle problem was overcome by an old friend of Albert Einstein's called Max Born. Exploring the results of the Schrödinger equation, Born tried a revolutionary assumption — and it worked perfectly. We are used to equations of motion, such as Newton's laws, telling us where a moving object will be after a certain amount of time. It had been assumed that Schrödinger's equation would do this. But Born realized that the equation was actually providing the *probability* of finding a particle in a particular location over time. It wasn't the particle that spread out; it was the chance of finding it in more distant locations.

This apparently small step totally transformed the nature of quantum physics. Quantum particles no longer continued to have exact positions when they were not observed. Instead, all that could be stated about them was the probability of finding them in a certain location.

▲ Max Born belatedly shared the 1954 Nobel Prize for Physics with Walther Bothe "for his fundamental research in quantum mechanics, especially for his statistical interpretation of the wave function."

Up to this point, physicists had thought of particles as being like the objects they saw around them. Just as the Ancient Greek atomists had considered atoms to be exactly the same stuff as the objects that are made from them, so particles were thought of as behaving like very small balls. However, the reality of quantum particles was very different. Instead, they became fuzzy patches of probability when not observed, as predicted by Schrödinger's equation. It was only when they interacted with something else that they once more had a specific location.

◀ With Born's assumption, a quantum particle no longer occupies a specific position but, over time, becomes a fuzzy collection of probabilities.

EINSTEIN THE COBBLER

Max Born and Albert Einstein exchanged many letters over the years, containing a fascinating mix of personal trivia and scientific discussion. In these letters we see the development of Einstein's doubts about the quantum theory that he had helped bring into being. And the aspect which became a sticking point for him was Born's introduction of probability.

It was in his letters to Born that Einstein wrote variants of his famous saying that God does not play dice, such as, "The theory says a lot, but does not really bring us any closer to the secret of the 'old one.' I, at any rate, am convinced that He is not playing at dice." Einstein believed that the universe operated on clear and specific principles. If Born's interpretation of Schrödinger's equation were true, particles did not have positions until they were observed, merely probabilities. Reality, Einstein believed, was not like this.

Einstein made it clear how he felt about this probabilistic aspect of quantum theory when he wrote to Born describing a quantum effect that appeared to be controlled by probability: "In that case, I would rather be a cobbler, or even an employee in a gaming house, than a physicist."

Einstein was not religious and seemed to use "God" to mean "the organizing principle of the universe."

"The whole thing is rather sloppily thought out, and for this I must respectfully clip your ear."

Even the great initial success of the quantum theory does not make me believe in the fundamental ce game."

"I cannot make a case for my attitude in physics which you would consider reasonable ... I cannot seriously believe in [quantum theory] because the theory cannot be reconciled with the idea that physics should represent a reality in time and space, free from spooky actions at a distance."

"Quantum mechanics is certainly imposing. But an inner voice tells me that it is not yet the real thing."

"My instinct for physics bristles at this. However, if one abandons the assumption that what exists in different parts of space has its own, independent, real existence then I simply cannot see what it is that physics is meant to describe."

"This theory reminds me a little of the system of delusions of an exceedingly intelligent paranoiac, concocted of incoherent elements of thoughts."

THE UNCERTAINTY PRINCIPLE

The probabilistic aspects of quantum theory led to one of its most famous results: Heisenberg's uncertainty principle. This has been widely misused to imply that everything is uncertain, but actually it tells us of a very precise relationship between pairs of properties of quantum particles.

The uncertainty principle shows that if we take one of these pairs for a particular quantum object — for example, position plus momentum, or energy plus time — then the more accurately we know one value, the less accurately we know the other. If, for example, we know exactly where a quantum particle is, its momentum could have any value; while if we know its momentum exactly, it could be anywhere.

Heisenberg first used the example of a microscope, where the photons used to view a particle push it off course. Bohr tore this apart, pointing out that the uncertainty principle does not require outside intervention.

▲ *Heisenberg in 1924.*

THE UNCERTAIN PHOTOGRAPHER

• • • • • • • • •

A useful analogy is the photographing of a fast-moving object. Photographed with a long exposure, the object blurs. This gives us a feel for how it is moving, but little idea where it is. With a short exposure, we get an exact location, but no feel for how it is moving. However, the principle is not about observation — it tells us about the actual nature of the particle.

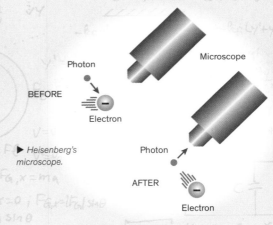

▶ Heisenberg's microscope.

Photon

Microscope

BEFORE

Electron

Photon

AFTER

Electron

▼ The uncertainty principle is like a photograph of a moving object, which can illustrate location or speed (but not both), depending on the exposure.

THAT CAT

If there is one image that is inevitably used to represent quantum weirdness, it is Schrödinger's cat — though in reality it adds little to our understanding of that weirdness.

A system that has a probability of being in more than one quantum state is said to be in a superposition of the states.

While Heisenberg was happy to go along with Born and put probability at the heart of reality, Schrödinger had more sympathy with Einstein's viewpoint. He dreamed up a thought experiment to illustrate his problem with the approach.

In the thought experiment, a cat is in a box with a vial of deadly poison. The vial can be broken by a mechanism

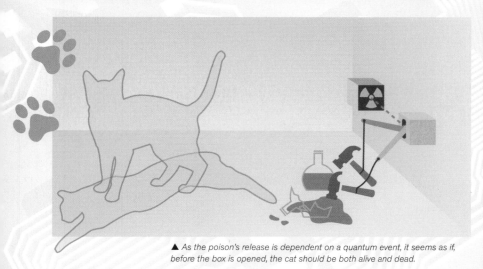

▲ As the poison's release is dependent on a quantum event, it seems as if, before the box is opened, the cat should be both alive and dead.

▶ *Life-size cat figure in the garden of Huttenstrasse 9, Zurich, where Erwin Schrödinger lived 1921–1926. Depending on the light conditions, the cat appears either alive or dead.*

triggered by the decay of a radioactive atom. Like many behaviors of quantum particles, nuclear decay is a probabilistic affair, and, according to Born's approach, if you leave the box for a while, the particle is neither whole nor decayed — there are just probabilities of it being in one state or the other. This implies that the cat is both alive and dead.

In practice, the experiment is meaning-less, as the interaction with the radiation detector is enough to ensure the atom will be in one state or another — but the cat lives on in the half-world of imagination.

UNREAL CATS

Scientists regularly overdramatize new experiments related to Schrödinger's cat — for example, in 2016 an experiment at Yale was described in the press as "Schrödinger's cat alive and dead even after you saw it in half." The reality was far more mundane. No cats were involved — just photons of light. These were set up in two boxes with different energies in each box. When linked, the boxes got into a "cat" state, where individual occupants of each box had both energies simultaneously. And this "superposed" state was maintained when the link between the boxes was severed — which was all that the "sawing in half" amounted to.

IN TWO PLACES AT ONCE?

The probabilistic interpretation of the Schrödinger equation explains why the double-slit experiment works, even when firing a single photon or electron at a time. The particles don't have a specific location so they have a probability of going through either slit, and the probability waves interfere to produce the pattern.

As a shorthand, the particle is often described as being in two places at once, so it "goes through both slits and interferes with itself." Media physicists often say this. This is understandable, because it's difficult to do anything else when you are trying to explain quantum physics in a soundbite.

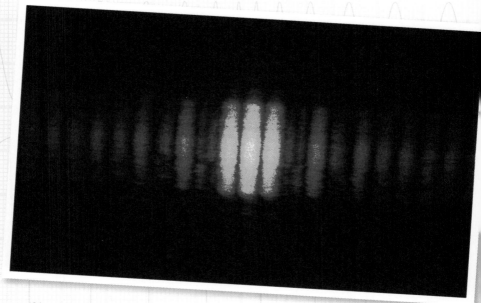

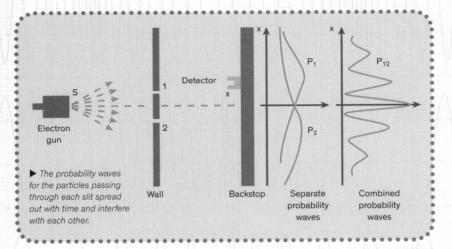

▶ The probability waves for the particles passing through each slit spread out with time and interfere with each other.

Electron gun

Detector

Wall

Backstop

Separate probability waves

Combined probability waves

However, the reality is stranger. The particle doesn't have a location until it interacts with something — in the case of Young's slits, when it hits the screen. All that exist are probabilities described by the Schrödinger wave equation. The particle doesn't fly in a straight line; it exists as an evolving set of probabilities until it hits the screen. It's not in two places at once. Indeed, it's not anywhere. Only the probabilities exist.

▶ Young thought that his slits made a certain observation of the nature of light, but quantum theory does away with certainty.

COPENHAGEN INTERPRETATION

Once Born's understanding of the Schrödinger equation became widely accepted, it was clear that quantum mechanics was very different from the clockwork physics of Newton and his successors.

If human observers were to get their heads around what was happening, they had to be either happy simply to accept, like Heisenberg, that the numbers matched reality, or they had to find a way to interpret what was observed. For many years, the dominant interpretation — one that is still common with some

modification — was the Copenhagen interpretation.

Emerging, as the name suggests, from the Danish physicist Niels Bohr, aided by Werner Heisenberg, the interpretation says that quantum entities left to their own devices are described by a probabilistic wave function, but when

◀ Bohr's institute in Copenhagen became the intellectual capital of quantum physics.

The Institute of Theoretical Physics, founded by Bohr at the University of Copenhagen in 1920, is now called the Niels Bohr Institute.

▼ In Newton's universe there is a simple, clockwork-like mechanical progress such that if we knew all the details of the current state of the universe we could predict what would happen forever in the future.

MYSTERIOUSLY COLLAPSING WAVES

• • • • • • • • • •

An essential component of the Copenhagen interpretation is "wave function collapse." Before a quantum system is observed, its properties — for example, its position or momentum — have a range of possible values with different probabilities. But when the system has been observed, the wave function, described by Schrödinger's equation, is reduced to a single value — the one that is measured. Some physicists are uncomfortable about their inability to explain *how* this happens, but many are happy simply to accept that this is an effective description of what is observed.

observed (when they interact with their surroundings), the wave function collapses to the specific observed value.

The interpretation also encompasses Heisenberg's uncertainty principle and the idea that it is inherently impossible to know anything other than probabilities about the "inner workings" of a quantum system.

PILOT WAVES

Although the Copenhagen interpretation satisfied many physicists who were happy simply to get on with their calculations and not worry about the inner workings of the quantum world, others thought that there had to be something more than probabilities involved.

One of the most persistent ideas is the pilot wave theory, which came out of de Broglie's work and was developed in detail by American-English physicist David Bohm. In this theory each quantum particle has a wave associated with it that guides it and evolves with time according to Schrödinger's equation.

This approach produces exactly the same outcomes as the Copenhagen interpretation, but its supporters are happier with a reality that does not depend on fuzzy probabilities.

The pilot wave theory still makes use of Schrödinger's equation, but this now describes a wave that somehow guides the particle, enabling it to undergo wave-like operations such as interference in Young's slits (see page 22), but still to be a particle. The initial version of the

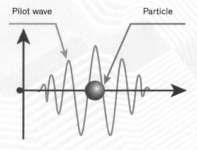

Pilot wave   Particle

▲ *In the pilot wave theory each quantum particle has a wave associated with it.*

theory, developed jointly by de Broglie and Bohm, could not deal with some of the behaviors of particles, but Bohm would later develop it to produce a better match with reality.

◀ *David Bohm was arrested in 1950 for refusing to testify before the House Un-American Activities Committee. He eventually abandoned the U.S. and settled in the UK.*

HIDDEN VARIABLES REVEALED

Unlike the Copenhagen interpretation, in which a particle does not have properties such as location until it interacts with something, the pilot wave theory has what are known as "hidden variables." These are actual values for, say, a particle's position at a particular time, but they are not accessible to the outside world.

DECOHERENCE

Although many physicists have no problem with the concept, some are uncomfortable with the idea of "wave function collapse," where the interaction of a particle with its environment causes it to go from being described as a collection of probabilities to having specific, measurable values.

Those who are uncomfortable with collapse don't like this sudden transition, nor the apparent distinction between the probabilistic behavior of quantum particles and the straightforward behavior of the physical objects around us, despite those objects being constructed of quantum particles.

To get around this, a different view called "decoherence," which doesn't involve collapse, is often now used. The idea of decoherence is that the wave function never collapses — when you look at the system as a whole, including, for example, the instruments doing the measuring, there is still a wave function describing the behavior of the quantum particles. But in their interaction with the outside world the quantum particles that are observed lose their "coherence," which can be thought of as their ability to act as independent entities described by separate wave functions.

Decoherence emerged from the many-worlds interpretation (see page 112), but is not dependent on it to be useful.

In effect, decoherence means that information is leaking from an isolated quantum system — a quantum particle, for example — into its environment. The quantum particle is no longer acting independently, but is linked to other particles around it, with the effect that it now appears to have a specific location.

MANY WORLDS

Another alternative to the Copenhagen interpretation is supported by a surprisingly large number of physicists, perhaps because it is the stuff of great science fiction. It is based on an idea originally set out in the PhD thesis of American physicist Hugh Everett.

In his "many-worlds interpretation," there is no collapse of the wave function. Instead, every possible outcome occurs, with the appropriate probability described by Schrödinger's equation. So, for example, in the two-slit experiment, there is no need for probability waves to go through both slits. Instead, the particle goes through each slit separately in a different quantum universe and the interference is between these different universes. In effect, every quantum possibility results in the emergence of a different universe across which, collectively, all possible outcomes occur.

◀ In the many-worlds interpretation, Schrödinger's cat is alive in one world and dead in the other.

Occam's razor is the work of the 13th- to 14th-century English friar, William of Ockham. Ockham is the correct spelling of the village in which he is believed to have been born, but the principle usually follows the Latinized version of its name.

SHAVING THE INTERPRETATION

Although there is no way to distinguish this kind of interpretation from variants of the Copenhagen interpretation, it is hard not to think that it fails the "Occam's razor" test, which states that the chosen theory should be the most parsimonious of those available.

QUANTUM TUNNELING

If quantum theory is correct, we can expect quantum particles such as atoms to behave very strangely.

Because its location is just a collection of probabilities that spreads out over time, it is possible for a quantum particle to find itself on the other side of a barrier without ever passing through it — as if a car suddenly jumped to the outside of a garage. As the possible locations spread out, eventually they will extend to the other side of the barrier, giving the particle a small but real chance of being on that other side of the barrier. Not only has this "quantum tunneling" been observed many times in experiments, it is used in electronics and our very existence depends on it.

Were it not for energy from the Sun, life on Earth would never have developed — and the Sun, like all stars, wouldn't work without quantum tunneling. This is because even the immense temperature and pressure at the heart of a star cannot push hydrogen nuclei sufficiently close together to overcome the repulsion produced by their positive

"Quantum tunneling" is a misnomer. The particle doesn't tunnel through a barrier; it simply appears on the other side.

charges and fuse. It is only because the nuclei have a small probability of tunneling through the barrier of the repulsion that fusion takes place.

Because the probability of tunneling is low, only a small percentage of the hydrogen nuclei succeed in tunneling through this barrier. However, there are such vast quantities of hydrogen in the Sun that several million tonnes of hydrogen manage to tunnel every second. Quantum tunneling effects contribute to biological processes such as photosynthesis and are used in some electronic devices.

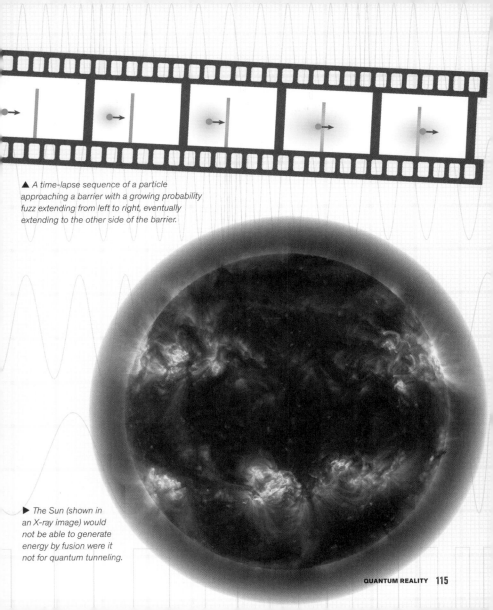

▲ A time-lapse sequence of a particle approaching a barrier with a growing probability fuzz extending from left to right, eventually extending to the other side of the barrier.

▶ The Sun (shown in an X-ray image) would not be able to generate energy by fusion were it not for quantum tunneling.

GOING SUPERLUMINAL

One of the more surprising implications of quantum tunneling is the ability to send information faster than the speed of light over very short distances, something that otherwise would be impossible according to Einstein's special theory of relativity, which doesn't allow faster-than-light communication.

This happens because the tunneling takes place instantly. This makes sense when you think about what is happening during tunneling — the particle isn't really traveling through the barrier; there is a probability that it is *already* at the other side.

▶ *In one superluminal experiment, microwaves tunnel across the gap between two large prisms when they should be contained within the first prism by the process of total internal reflection.*

In superluminal (faster-than-light) experiments, particles such as photons of light are fired at a barrier. Most bounce back, but a few tunnel through. As they spend zero time in the barrier, their overall speed across the whole apparatus is greater than the speed of light. Imagine, for example, an apparatus where light covered a distance of a millimeter in the usual fashion, then tunneled across a millimeter gap instantly. The result would be that photons had covered 2 millimeters in the time light takes to cross 1 millimeter — they would have traveled at twice light speed. However, the effect is so small that it is impossible to do anything with the apparently time-bending possibilities of faster-than-light signaling.

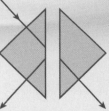

◀ Quantum particles should reflect within the first prism, and most do, but some tunnel across the gap, producing a faster-than-light passage.

After being told that no information could be sent through a superluminal experiment, German professor Günter Nimtz demonstrated the effect with a recording of Mozart's Symphony No. 40 that has traveled at four times the speed of light: visit http://universeinsideyou.com/experiment7.html.

CHAPTER 5

QUANTUM ELECTRODYNAMICS (QED)

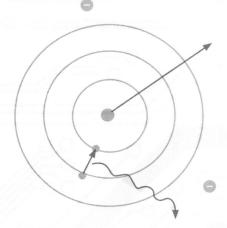

▶ *The ebullient American physicist Richard Feynman would become the figurehead for QED — quantum electrodynamics — the science of light and matter.*

PAULI'S ESCAPE CLAUSE

Niels Bohr used the early ideas of quantum physics to build a model of the hydrogen atom that seemed to work perfectly, but when he tried to extend it to cover the other elements there were problems (see page 76).

In Bohr's atom there was simply a set of orbits for the electron to jump between. But with the heavier elements, there were extra spectral lines that could be explained only by a more complex structure. Rather than having a single number — effectively the number of allowed orbits — experimental evidence implied that four different parameters were required to establish the energy of the orbits an electron could occupy: they would be called "quantum numbers."

The Austrian physicist Wolfgang Pauli suggested that no two electrons in the same atom could have all four quantum numbers the same as each other.

EXCLUSIVE ATOMS

· · · · · · · · · ·

This "Pauli exclusion principle" explains why atoms with many electrons don't end up with all the electrons in the same orbit, even at their lowest level of energy. The number of electrons left in the outer orbit or "shell" establishes the chemical properties of that element.

▶ *Wolfgang Pauli won the 1945 Nobel Prize for Physics for "the discovery of the Exclusion Principle, also called the Pauli Principle."*

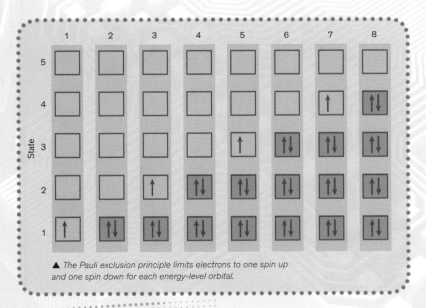

▲ The Pauli exclusion principle limits electrons to one spin up and one spin down for each energy-level orbital.

Pauli is nearly as well known in psychoanalytic circles as in physics. After he suffered a breakdown he became one of Carl Jung's patients, but was soon helping Jung develop his theories.

▼ The emission spectrum of iron has an extremely large number of lines, each from a different electron transition, implying a more complex electronic structure than Bohr's model provided.

DIRAC AND RELATIVITY

The last of the great names in the first cohort of quantum revolutionaries is the English physicist Paul Dirac.

By the time that Schrödinger's equation had become well established, physicists were very familiar with another of Einstein's contributions from 1905 — the special theory of relativity. This showed that space and time were intimately linked, resulting in a need to revise Newton's laws of motion. When an object moved very quickly, at close to the speed of light, quantities such as distance and momentum had to be modified,

▲ Paul Dirac shared the 1933 Nobel Prize for Physics with Schrödinger for "the discovery of new productive forms of atomic theory."

▼ Dirac developed his equation in his rooms in St John's College, Cambridge.

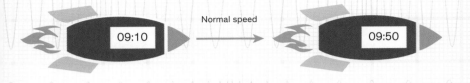

Normal speed

09:10 → 09:50

Near light speed

09:10 → 09:16

▲ Spacecraft moving at normal speed and at near light speed. The latter has a clock running much slower, is squashed in the direction of travel and has a much higher mass.

◄ Special relativity shows that time, momentum and more are influenced by relative motion.

depending on the relative motion of the object and an observer.

Although Schrödinger's equation was at the heart of the new 20th-century physics, it was a "classical" equation in that it used Newtonian laws of motion, rather than those modified by special relativity. This wasn't a problem when dealing with a slow-moving particle — but electrons, for example, often move at near the speed of light. Working alone in his Cambridge rooms, Dirac wrestled into shape a new equation for the electron that added relativity into the mix. However, to do so he had to make a strange assumption.

Bristol-born Dirac's father was French and spoke to him only in French, while Dirac's mother spoke only English. When young, Dirac believed that men and women spoke different languages.

123

THE INFINITE SEA

Dirac's equation did a great job of predicting the behavior of the electron, but it also stated that electrons could have both positive and negative energy levels. If this were the case, there was nothing to stop electrons taking quantum leaps into lower and lower negative levels, giving off infinite quantities of energy. Yet this was not borne out by experiments — every electron that had ever been observed had positive energy.

A lesser physicist might have dropped his equation, but Dirac was sure it was correct and looked for a way out. He suggested that the universe contained an infinitely deep sea of negative-energy electrons, filling all the possible spaces. Because of Pauli's exclusion principle, this meant an electron could never plunge down into negative energy.

◄ In the Large Hadron Collider at the European Organization for Nuclear Research (more commonly known as CERN) in Geneva, Switzerland, large numbers of particles are created from energetic collisions.

$$\left(\beta mc^2 + c \left(\sum_{n=1}^{3} \alpha_n p_n \right) \right) \psi(x, t) = ih \frac{\partial \psi(x, t)}{\partial t}$$

▲ The Dirac equation.

POSITIVE HOLES

· · · · · · · ·

There was one exception to the full sea. Occasionally a negative-energy electron would be boosted to positive energy by incoming light, leaving behind a hole that could be filled by a normal electron dropping into it. This presented a way of experimentally testing Dirac's concept.

Dirac was infamously lacking in social graces. When someone said, in a question-and-answer session after a Dirac lecture, "I don't understand the equation in the top-right-hand corner of the blackboard," Dirac said nothing, resulting in a long, uncomfortable pause. Eventually Dirac replied, "That was not a question, it was a comment."

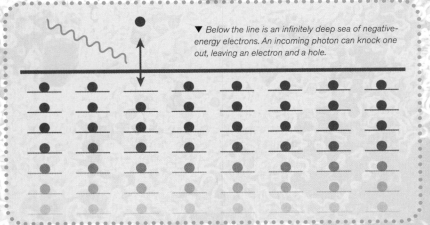

▼ Below the line is an infinitely deep sea of negative-energy electrons. An incoming photon can knock one out, leaving an electron and a hole.

◄ The first positron to be discovered, using a cloud chamber in which particles leave a trail of water droplets. The curved lines show an electron and a positron, bent in opposite directions by a magnetic field.

MISSING THE EVIDENCE

• • • • • • • • • • •

Ironically, Dirac missed the first evidence supporting his theory. In 1931, American physicist Carl Anderson made the first detection of a positron in cosmic rays — high-energy streams of particles that reach Earth from space. Anderson's work was presented in a seminar in Cambridge, but Dirac was on sabbatical in America at the time.

ENTER THE POSITRON

Although looking for a hole in an infinite sea might have seemed a challenge, Dirac realized that a missing negative-energy electron would appear identical to a detectable positive-energy particle. He initially thought this was the proton, but soon realized his positively charged particle needed to have the same mass as the electron. This was the first suggestion of the existence of antimatter — and the antielectron, or positron, should be detectable because it would have the same mass as an electron but would be bent in the opposite direction by an electromagnetic field.

Although the need for the infinite sea would be removed by quantum field theory (see page 128), Dirac's equation and antimatter had a lasting legacy.

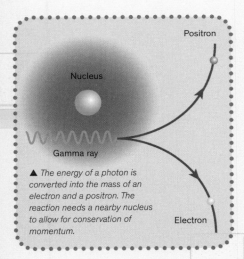

▲ The energy of a photon is converted into the mass of an electron and a positron. The reaction needs a nearby nucleus to allow for conservation of momentum.

Carl Anderson was the PhD student of the American physicist Robert Millikan, who had proved that Einstein was right about the photoelectric effect (see page 74). It was Millikan who gave the seminar that Dirac missed.

The first antimatter particles: each matter particle has an equivalent antimatter particle, with the same mass but differing in one or more of its properties.

Matter	Electron	Proton	Neutron
Mass (kg)	9.109×10^{-31}	1.673×10^{-27}	1.675×10^{-27}
Charge (e)	−1	1	0
Magnetic moment pN	−1.001	2.793	−1.913

Antimatter	Antielectron (positron)	Antiproton	Antineutron
Mass (kg)	9.109×10^{-31}	1.673×10^{-27}	1.675×10^{-27}
Charge (e)	1	−1	0
Magnetic moment pN	1.001	−2.793	1.913

FIELDS EVERYWHERE

Until the 1930s, quantum physics had treated the phenomena it studied as waves or particles, but now a third way to approach them mathematically was thrown into the mix — fields.

The idea of a field had been introduced by Michael Faraday in the 19th century. It was a bit like a contour map that showed the level of something at every point in space, but operating in three dimensions of space and one of time.

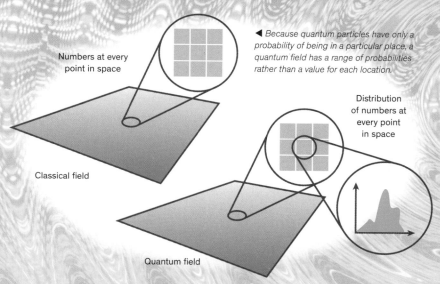

◀ Because quantum particles have only a probability of being in a particular place, a quantum field has a range of probabilities rather than a value for each location.

Numbers at every point in space

Classical field

Distribution of numbers at every point in space

Quantum field

Quantum field theory imagined the whole of the universe permeated by a number of fields — for example, the electromagnetic field. Rather than light being treated as a wave or a particle, it could be considered as a traveling fluctuation in the electromagnetic field. This proved a hugely useful breakthrough in the ability to predict how quantum phenomena will behave.

FIELDS ARE MODELS

It should be stressed that light, or matter, is not particles or waves or disturbances in a field. Each of these is a model to help us understand and predict the behavior of quantum phenomena. But in reality, light, for example, is just light.

▶ The contours on a map link points with the same height. In effect this is the "height field."

QUANTUM ELECTRODYNAMICS

The work started by Paul Dirac would be developed independently by a trio of scientists — Richard Feynman and Julian Schwinger in the United States and Shinichiro Tomonaga in Japan — into a working theory of the interaction of light and matter, known as quantum electrodynamics or QED.

Although strictly a field theory, QED can be best envisaged (without delving into the mathematics) as a theory that explains all the interactions of light with matter, and matter with matter — which amounts to most of our experience of the world around us — using quantum particles.

Richard Feynman said in an introductory lecture on QED, "I want to emphasize that light comes in this form — particles. It is very important to know that light behaves as particles, especially for those of you who have gone to school, where you were probably taught something about light behaving like waves. I'm telling the way it *does* behave — like particles."

▲ *Feynman, Schwinger and Tomonaga won the 1965 Nobel Prize for Physics for QED.*

QED is an incredibly effective theory, producing values that are so close to observation that Feynman said it was the equivalent of calculating the distance from New York to Los Angeles to the width of a human hair.

Feynman became famous for revealing the cause of the Challenger space shuttle disaster by plunging a section of rubber O-ring into his glass of iced water at a televised hearing and showing how it lost flexibility at low temperatures.

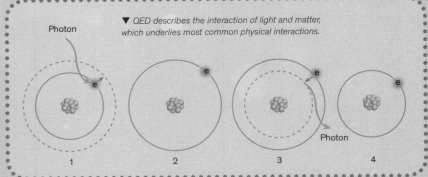

▼ *QED describes the interaction of light and matter, which underlies most common physical interactions.*

Photon

e

e

e

e

Photon

1   2   3   4

FEYNMAN'S VAN

Of the three physicists who won the Nobel Prize for QED, Richard Feynman was both the most charismatic and the best at communicating quantum physics to a wider audience. This emphasis on communication is evident in his development of the Feynman diagram. This is a way of representing a quantum interaction as a series of lines on a plot of space and time.

Not only would Feynman diagrams prove useful as a way of exploring what is happening when quantum particles interact, they would even have a role to play in the calculations that usually accompany quantum physics, because the diagrams are visual versions of mathematical formulas. The probabilities of a particular outcome can be calculated as the sum of the outcomes of the different possible Feynman diagrams. (In practice not every diagram is drawn; they are usually added only until the contribution is small enough to be ignored.)

The spidery diagrams became such a Feynman trademark that he had a van decorated with them, which he proudly drove around the California Institute of Technology (Caltech) campus.

Feynman was not impressed by "sophisticated" culture, but he was an enthusiastic player of the bongos.

◀ Richard Feynman was a world-renowned speaker — his collected lectures on physics, the so-called "red books," have become a classic in their field.

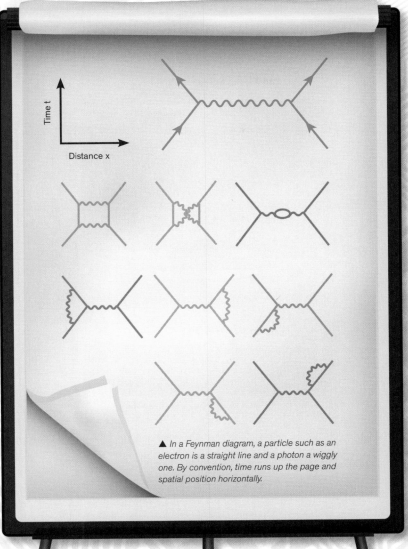

▲ In a Feynman diagram, a particle such as an electron is a straight line and a photon a wiggly one. By convention, time runs up the page and spatial position horizontally.

133

▲ *Billiard balls have straight-line, symmetrical paths.*

TAKING EVERY PATH

The way that Feynman diagrams play a role in QED calculations is mind-blowing, though the method is simple. For example, suppose two particles come together and scatter, bouncing off each other due to electromagnetic repulsion. Classical physics represented this as two billiard balls traveling in straight lines and bouncing off each other at the same angle. But the quantum world is very different.

The "billiard ball" picture

gives one Feynman diagram — but there are many more in which, for example, the angles are not symmetrical, or one of the particles makes a detour around the Sun and then collides with the other. In principle, *every* possible path is taken. However, some of these paths are of such low probability that they can be ignored. Others cancel each other out. And the final result is the expected billiard-ball-like behavior. Clearly, it is impossible to draw every possible Feynman diagram, but by sticking to high-probability paths, the outcome can be deduced.

THE MAGIC MIRROR

The reality of the strange paths can be demonstrated with a mirror and a beam of light. We expect light, as we are taught at school, to reflect off a mirror at the same angle as it arrives. But this is only because the photons of light have a property called "phase" that is a little like each photon carrying a clock with fast-turning hands. If the hands of

A NECESSARY COMPLICATION

If everything other than the expected behavior is eliminated, why bring in the extra paths in the first place? It very soon becomes clear that these "unnecessary" paths do exist — and they explain how quantum particles behave in the strange ways that they do.

Feynman diagrams required a total rethink of the traditional approach to optics.

the clocks line up at the same point in space and time, the particles are in phase and reinforce each other. If they are out of phase, they cancel each other out (phase is how particles produce wavelike behavior).

All the particles reflecting at strange angles have phases that cancel out, so the result is the expected classical reflection. But if you remove strips of the right width from the mirror, the result is that photons reflect at a totally unexpected angle, because the phases at this new angle are no longer canceling out. As the rate of phase change is related to frequency and hence color, different colors reflect off such a mirror at different angles.

The rainbow seen when looking at a CD or DVD at an angle is caused by this effect, with the pits in the disc acting as the removed strips.

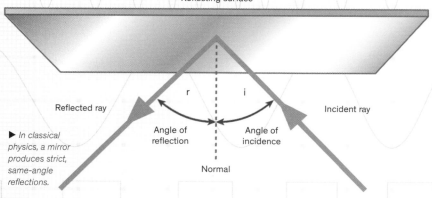

Reflecting surface

r   i

Reflected ray   Incident ray

Angle of reflection   Angle of incidence

▶ *In classical physics, a mirror produces strict, same-angle reflections.*

Normal

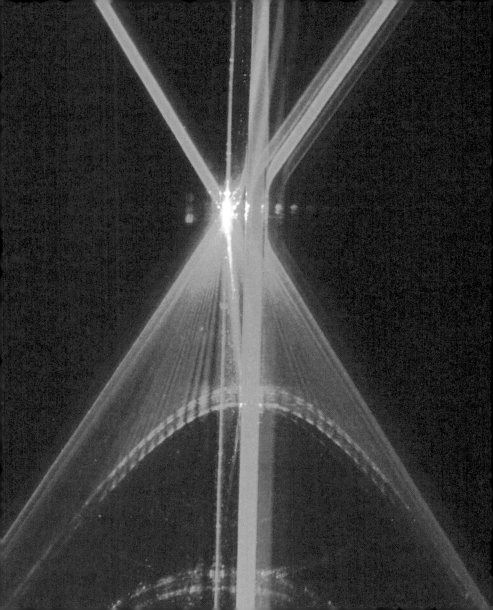

QUANTUM LENSES

Feynman's work on QED showed that all
the behaviors of light that needed a wave
description could be explained using a
particle model. A good example is a lens,
which seems to have odd behavior when
considering particles.

If there is no lens, the phases of all
the longer paths cancel each other out,
leaving a beam of light that travels by the

shortest route — a straight line from A to
B. (Note that the paths don't have to be
straight lines, but random wiggly paths
tend to cancel each other out.)

The reason the non-straight paths
cancel out is that their phases differ,
because the light has traveled for

different times. A lens places different thicknesses of glass in the way of the paths, and because light travels more slowly through glass than through air, more glass means a longer travel time. The result is that light takes the same time to travel all the paths — all the light arrives in phase and adds up rather than cancels out.

The word "lens" is taken from the Latin for "lentil" because the shape of a convex lens is reminiscent of the popular pulse.

RENORMALIZATION

Though QED proved remarkably effective, it also presented a problem. When the different probabilities for some quantum measurements were added together, they ran off to infinity. This can happen with the electric field of an electron. The strength of the field is proportional to the inverse square of the distance from the center of the particle. As the electron has zero diameter, its own electrical field acting on itself, getting exponentially bigger as you get closer to its center, shoots off to infinity.

Faced with a theory that worked extremely well in many respects but sometimes ran away with itself, QED theorists resorted to a fudge, which became known as "renormalization." Effectively, they substituted correct values from observation for these runaway figures. Only relatively few values had to be manually added: renormalization still left QED with many predictions that matched experiments better than any other theory.

A SERIES PARALLEL

• • • • • • • • • •

The unexpected infinities are a bit like the sums of two very similar-looking series of numbers. If you add up this infinitely long series of fractions:

$$1 + \frac{1}{2} + \frac{1}{4} + \frac{1}{8} \dots$$

the total is 2.

However, sum this infinite series:

$$1 + \frac{1}{2} + \frac{1}{3} + \frac{1}{4} \dots$$

and the total is infinity.

When physicists encounter infinity, they generally believe that it shows a limitation in their theories — though we still don't know whether the universe is finite or infinite.

▲ In a black hole, all the matter in a star collapses to a point with infinite density. Although we speak of black holes as existing, in reality this infinite value means that our current physics is no longer applicable.

POLARIZATION

An important quantum property of a photon is its polarization. Just as phase had shown itself in wavelike behavior, scientists were aware of polarization long before quantum physics was developed.

The polarization of a photon means effectively that a particular direction is associated with it. (The quantum details are more complex, requiring four values, as polarization can change with time, but it's easiest to think of it as a direction.) Usually a beam of light contains photons with random polarizations, but some

▲ *Iceland spar, a transparent form of the mineral calcite, bends light to different degrees depending on its polarisation. The result is the multiple image seen*

materials act rather like optical slots through which only photons polarized in a particular direction can pass.

Polarization is a true quantum property. If we send photons through three polarizing filters arranged as in the diagram below, some will make it through, even though they have passed through filters at 90 degrees to each other. This is because the middle filter produces photons that are in a superposed state of both horizontal and vertical at the same time.

Engineer Edwin Land dropped out of Harvard aged 18, fascinated by the phenomenon of polarization. In his garage laboratory he developed a plastic sheet with tiny polarizing crystals embedded in it, which became known as Polaroid.

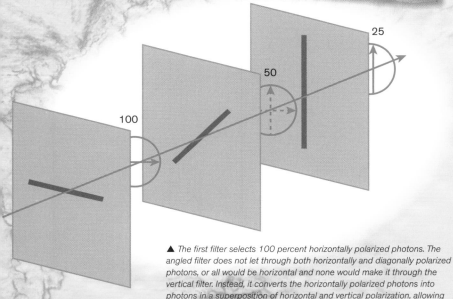

▲ The first filter selects 100 percent horizontally polarized photons. The angled filter does not let through both horizontally and diagonally polarized photons, or all would be horizontal and none would make it through the vertical filter. Instead, it converts the horizontally polarized photons into photons in a superposition of horizontal and vertical polarization, allowing some to pass through the final filter.

TIME TRAVELERS

The quantum particles mapped out by Feynman diagrams are not only free to take any spatial path, but can also be flexible in their route through time.

A common quantum interaction is for an electron to absorb a photon of light and then give one off a little later. This process is responsible for everything from the blue sky we see above us to the reflection of light off objects. But there is nothing in the "rules" to say that an electron cannot emit a photon, travel backward in time, absorb a photon, then carry on forward in time again.

An electron traveling backward in time turns out to be physically identical to a positron traveling forward in time. So this apparently impossible scenario can also be looked at as the energy of a photon generating an electron/positron pair. The newly generated positron — the antimatter equivalent of the electron that Dirac had predicted (see page 124) — combines with the original electron to produce a new photon, while the electron from the pair carries on. So we begin and end with a photon and an electron.

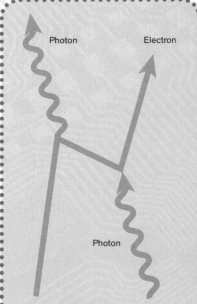

▲ A positron can be represented as an electron traveling backward in time, shown in this Feynman diagram as the central straight line.

When light is "reflected" off an object it is actually absorbed, boosting an electron to a higher energy orbit, then reemitted.

Most physicists, when pressed, are likely to admit that they don't believe that a positron truly is an electron traveling backward in time. However, for the calculations that lie behind the Feynman diagrams it can often be useful to work with this picture, much as imaginary numbers are used with Schrödinger's equation.

▼ *Time travel as portrayed in science fiction may not be possible, but some time-travel-like features are inherent to physics.*

ANTIPARTICLE REALITY

• • • • • • • • • • • • •

All antiparticles can be represented on Feynman diagrams as "normal" particles traveling backward in time. Feynman suggested that the backward-in-time electron was just as real as the positron — but because we experience the world moving forward in time, we observe only the positron.

ADVANCED AND RETARDED LIGHT

Feynman was also responsible for another time-bending way of looking at a quantum event. Maxwell's equations for electromagnetism had two solutions describing two different types of electromagnetic wave: retarded waves, which are the ones we experience; and advanced waves, which travel backward in time from target to source. The advanced wave solution had simply been ignored because it seemed to have no connection to reality.

Along with his former doctoral supervisor, the American physicist John Wheeler, Feynman suggested that advanced waves really exist.

This overcomes a potential problem when an electron emits a photon — the recoil produced on the electron by the conservation of momentum (just like

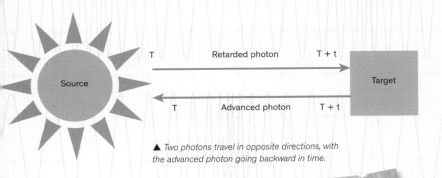

T   Retarded photon   T + t

Source   Target

T   Advanced photon   T + t

▲ *Two photons travel in opposite directions, with the advanced photon going backward in time.*

the recoil of a gun when it fires a bullet) should produce a kind of feedback effect on the electron that results in infinite energy levels, an effect that clearly does not happen in practice. Wheeler and Feynman suggested that *two* photons were produced: the advanced photon travels backward in time from the destination of retarded photon to the electron that caused the recoil, removing the feedback.

Each photon would have half the energy of the observed one, and because of their opposing motion through time as well as space they would always be in the same place at the same time. The outcome cannot be distinguished from that produced by a single conventional photon — but it makes sense of Maxwell's equation and does away with the inconvenient infinite value.

Like Feynman, John Wheeler was a huge character. He is often said to have coined the term "black hole," but in fact he only popularized it after it was first used by an unknown speaker at a conference.

In principle, advanced waves provide a complex mechanism for sending a message backward in time, but to make it work we would need to find a region of space lacking in absorbers of light and to have the ability to manipulate an absorbing material into and out of a light beam at a great distance.

THE AMPLITUHEDRON

Feynman diagrams can also be used for complex quantum particle interactions, such as QCD — quantum chromodynamics — which describes the interaction of quarks and gluons, the fundamental particles that make up protons and neutrons (see Chapter 7). In practice, however, the interactions are so complex that the number of diagrams that must be considered becomes unmanageable.

Although the concept is still under development, it's possible that a replacement diagram for more complex interactions is an amplituhedron. This is based on a mathematical structure called the positive Grassmannian, which describes the space inside a triangle. When this was expanded to multiple dimensions to encompass the space described by intersecting planes, it produced a kind of hyper-Feynman diagram that can deal with multiple aspects of the interaction of particles at once.

▼ A step on the way to the development of the amplituhedron was the production of networks of particle interactions in a special type of space–time geometry known as "twistor space."

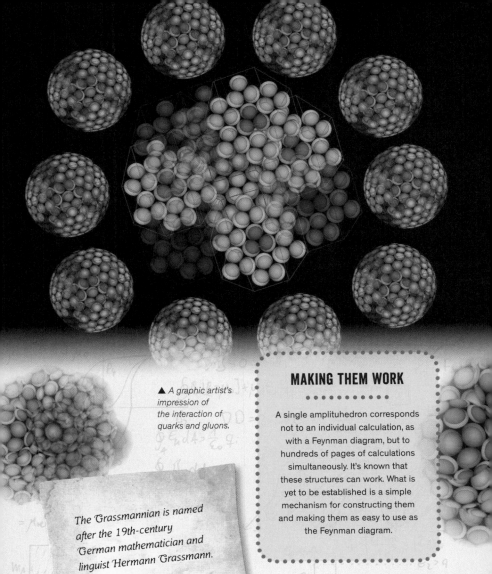

▲ *A graphic artist's impression of the interaction of quarks and gluons.*

The Grassmannian is named after the 19th-century German mathematician and linguist Hermann Grassmann.

MAKING THEM WORK

• • • • • • • •

A single amplituhedron corresponds not to an individual calculation, as with a Feynman diagram, but to hundreds of pages of calculations simultaneously. It's known that these structures can work. What is yet to be established is a simple mechanism for constructing them and making them as easy to use as the Feynman diagram.

CHAPTER 6

A TANGLED WEB

▶ Quantum entanglement, perhaps the strangest phenomenon in all of physics, shows that particles can be linked to each other across any distance.

QUANTUM SPIN

QED made understanding the quantum world a little more straightforward, but there were still some aspects of quantum physics that were considered downright weird. This chapter focuses on quantum entanglement, a feature of the quantum world that Einstein described as "spooky."

▲ *Familiar objects can spin around an axis in any direction.*

To understand the peculiarities of entanglement we need to get a feel for one of the properties of quantum particles — spin. It seems a straightforward concept. We are used to things spinning around, whether it's the Earth rotating or the spin put on a tennis ball. And quantum spin was so called because it was thought to be analogous to the familiar phenomenon. But, as is so often the case in the quantum world, things are very different.

When you measure the direction of spin of a quantum particle, it comes out as either up or down. Before making the measurement, a particle is in a "superposition" of these states — rather than having a specific value, all that exists is the probability. So, for example, in a selected direction of measurement, it might have a 40 percent chance of being "up" and a 60 percent chance of being "down."

The superposition sounds a little like our lack of knowledge of a tossed coin before we look at it. We say that it has a 50:50 chance of being heads or tails. In reality, the coin has a single value; we just don't know what that value is. In a superposed quantum state, only the probabilities exist. The particle is not in either state until the measurement is made, when it takes on either "up" or "down," but we can discover the probabilities of each possible outcome.

Quantum spin is most often measured using a so-called Stern–Gerlach apparatus, named after the German physicists

'Einstein called entanglement "spukhafte Fernwirkung," which roughly translates as "spooky action at a distance."

Otto Stern and Walter Gerlach, who devised the experiment in 1922. It involves passing the particles through a pair of specially shaped magnets, producing an inhomogeneous magnetic field that splits off particles with opposing spins.

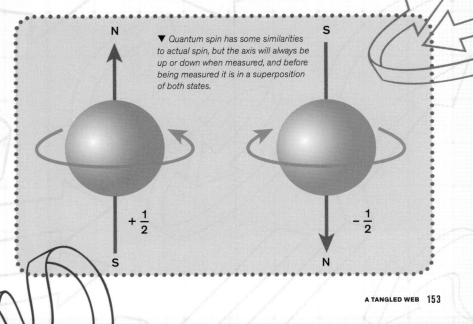

▼ *Quantum spin has some similarities to actual spin, but the axis will always be up or down when measured, and before being measured it is in a superposition of both states.*

$+\frac{1}{2}$

$-\frac{1}{2}$

THE MYSTERY OF WINDOWS

Another important contributor to the entanglement story is something clearly demonstrated when considering the window of a lit-up room at night. From inside, you see a reflection of the room in the glass.

But go outside and you can see into the room. Some light from the room is reflecting back from the glass, while most of it passes straight through to the outside. When a photon arrives at the surface of the glass, how does it know whether to reflect or pass through? Even stranger, the thickness of the glass

▲ *Some of the light from the room is reflected back, while the rest passes through the glass — all purely dependent on probability.*

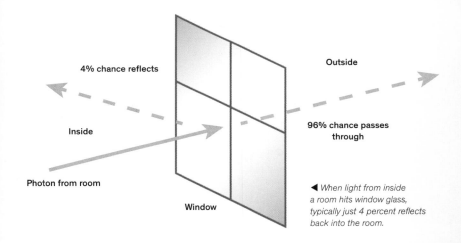

4% chance reflects

Outside

Inside

96% chance passes through

Photon from room

Window

◀ *When light from inside a room hits window glass, typically just 4 percent reflects back into the room.*

influences how likely a photon is to reflect back from the *inner* surface, even though the photon apparently has no way of knowing at this point how thick the glass is.

Newton, who thought light was a stream of particles, was incapable of explaining the selective reflection. He first thought it was due to scratches on the glass — but polishing didn't weaken the effect. We now know that it's a quantum effect. A photon's "decision" to reflect or pass through is not about the state of the glass but purely dependent on probability.

The thickness of the glass is significant because the photon's probability wave extends to the back of the sheet of glass, which means that whether it passes through or not is dependent on the thickness.

The same amount of light reflects back into the room during the daytime, but it is washed out by the much stronger light from outside.

BEAM SPLITTERS

The glass in a typical window, which reflects some light and lets the rest through, is a simple example of a frequently used quantum physics device called a "beam splitter." A sheet of glass is not very useful as a beam splitter in experiments, because most of the light passes through; an inefficient mirror is a more effective beam splitter.

These often feature in police stations in crime dramas, where they tend to be called "two-way mirrors" — an odd term, because they act as mirrors when viewed from one side only, while appearing to be see-through glass when viewed from the other. Such two-way mirrors typically reflect most of the light to conceal their nature, requiring low lighting conditions in the viewing

The correct term for a "two-way mirror" is "half-silvered mirror." The reflective coating (usually aluminum rather than silver) is sufficiently thin to let some of the light through.

◀ On hitting a half-silvered mirror, roughly half of the photons reflect, while half pass through.

▲ *Two-way mirror in police station.*

room. But for a scientific experiment, the ideal is that the mirror reflects half of the incoming light. More sophisticated beam splitters, often now used in quantum experiments, are made of pairs of prisms glued together, which have a similar effect.

SPLITTING FOR ENTANGLEMENT

Beam splitters provide a mechanism (in a setup requiring a pair of splitters) to get two photons to interact and enter the special quantum state called entanglement, producing remarkable properties such as the ability to interact instantly at great distances.

EINSTEIN'S CHALLENGES

Because of his distaste for the probabilistic aspects of quantum physics, Einstein dreamed up a series of problems that seemed to show that the theory was faulty. He presented these challenges to Niels Bohr at conferences, typically over breakfast. Bohr would go away and think, before coming back to show that Einstein had missed some aspect of the problem that meant quantum theory was sound.

After Bohr repelled him throughout the 1920s, Einstein abandoned his attack until 1935, when he devised his final assault, using the bizarre nature of quantum entanglement.

▲ Einstein and Bohr regularly debated problems arising from quantum physics.

GENERAL THEORY OF RELATIVITY

Einstein's general theory of relativity describes the impact of matter on space and time. It predicts that matter produces a warp or curve in space–time, resulting in the effects of gravity. One of the implications of the general theory is that time runs more slowly under stronger gravitational fields, so as you move away from a planet, a clock will start to run faster.

The challenge that Bohr found most difficult involved a clock in a box, which emitted a photon of light at a specific time, apparently allowing both energy and time to be measured exactly, which challenged the uncertainty principle. Therefore, Einstein thought that the uncertainty principle (see page 100) did not

hold here, which put the whole basis of quantum theory at risk. Bohr struggled all day with this problem before realizing that, ironically, Einstein had missed an implication of his own general theory of relativity. The thought experiment relied on the clock making a small upward movement, as the release of the photon reduces the energy in the box, and energy, like matter, is influenced by gravity. According to relativity, this meant that the clock experienced slightly less gravity and ran faster. This effect produces an inaccuracy in the time measurement that offsets the problem raised by the thought experiment.

Bohr would always regard Einstein as a thorn in the side of his theories. His colleague Abraham Pais recounted listening to Bohr work through a problem while occupying an office adjacent to Einstein's at the Institute for Advanced Study. Bohr was muttering "Einstein" to himself repeatedly as he stared out of the window.

At this point, Einstein crept into the room. He had been told not to buy tobacco by his doctor, so had decided to borrow Bohr's. Einstein tiptoed to the desk to help himself as Bohr continued to mutter "Einstein." As Einstein reached the desk, Bohr came out with a final, firm "Einstein!" and turned to find the subject of his thoughts standing right in front of him. "There they were," Pais said, "face to face, as if Bohr had summoned him forth. It is an understatement to say that for a moment Bohr was speechless."

When Einstein came up with the clock challenge, an observer noted that he walked quietly away from the meeting with a "somewhat ironical smile," while Bohr "trotted excitedly beside him."

EPR

In 1935, Einstein joined two young physicists, the Russian-born American Boris Podolsky and the American-Israeli Nathan Rosen, to write a paper showing that quantum theory was ridiculous, as it predicted the improbable implications of quantum entanglement. The paper is universally known, after the initials of its authors, as "EPR."

Like Einstein's earlier challenges, this too involved a thought experiment. The original paper was unnecessarily complicated by using two different quantum properties, but in a simplified form it involved producing two quantum particles in the entangled state from a single original. These particles traveled in opposite directions to a great distance.

The (imaginary) experimenter then measured the spin of one particle and found it to be up. Until the measurement was taken, each particle was in a superposed state, up and down, with a 50 percent probability of either. However, because spin is a property that is conserved in physics, and the original particle had no spin, the second particle now immediately had to become spin down in order to balance the first particle. This, EPR suggested, was entirely unreasonable, as the particles would need to communicate their states of spin to each other instantly, however far apart they were — but relativity limited communication to the speed of light.

▼ In the EPR experiment, when two entangled particles are at a distance from each other, observing one has an instant effect on the other.

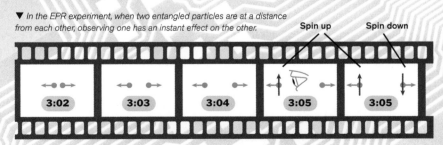

Spin up   Spin down

3:02   3:03   3:04   3:05   3:05

(left to right) Einstein, Podolsky and Rosen.

Einstein thought the EPR paper's initial use of two properties was unnecessarily messy. He later said of this complication, "Ist mir Wurst," literally "It is sausage to me," meaning "I couldn't care less."

EPR's actual title is "Can Quantum-Mechanical Description of Physical Reality Be Considered Complete?" It concludes, "While we have thus shown the wave function does not provide a complete description of the physical reality, we left open the question of whether or not such a description exists. We believe, however, that such a theory is possible." By "such a theory" the authors meant one where it was possible to know everything about quantum particles, rather than always being limited to knowing probabilities before a measurement was taken — as quantum theory required.

LOCALITY

Einstein didn't make EPR an out-and-out assault on quantum physics, but instead offered two possible conclusions. He had previously argued that quantum particles, like those in the EPR experiment, were not in a 50:50 probability state, but knew all along whether they would be spin up or spin down: this information had to be stored somewhere inaccessible — in so-called "hidden variables" (see page 109).

A later physicist likened it to a Dr. Bertlmann, who always wore different-colored socks. If you saw a green sock on one of his feet, you instantly knew the other sock wasn't green. The information was there, but hidden. EPR concluded that either quantum physics was wrong and there were such hidden variables, or locality had to be forgotten.

◀ We can deduce that Dr. Bertlmann's other sock is not green without ever seeing it.

Locality means that something cannot influence something else at a distance without communication between them. Locality was the reason that Newton's ideas on gravity were doubted by his contemporaries, who mocked the idea that the Earth could be "attracted" to the Moon at a distance, calling it an occult effect. Physicists breathed a sigh of relief when general relativity provided a mechanism for gravity that required no "action at a distance." The EPR paper ponders disposing of locality but then concludes, "No reasonable definition of reality could be expected to permit this."

The Northern Irish physicist John Bell wrote a paper entitled "Bertlmann's Socks and the Nature of Reality" for the Journal de Physique in 1981.

▲ In nature, something travels from A to B to cause action at a distance.

NO ACTION AT A DISTANCE

Action at a distance, such as the ability of two entangled particles to interact with each other instantly, is never normally seen in physics. When we hear something across a room, for instance, sound waves move through the intervening air. When the apparent action at a distance of magnetism pulls a piece of metal toward a magnet there is a flow of photons in between, carrying the electromagnetic force. But entanglement has no equivalent.

BELL'S INEQUALITY

The response to EPR was generally to shrug and ignore it. Niels Bohr always claimed he did not get the point of it. And as quantum theory was increasingly shown to represent reality with remarkable accuracy, no one took much notice. But in the 1960s, John Bell, a physicist from Northern Ireland working at the European Organization for Nuclear Research (more commonly known as CERN) near Geneva, Switzerland, returned to the paper.

It was Bell who told the story of his friend Dr. Bertlmann's socks, and he very much sided with Einstein in being uncomfortable with the probabilistic

Bell once said of the role of probability in quantum physics: "I hesitated to think it might be wrong, but I knew that it was rotten."

nature of quantum physics. No one had carried out the EPR thought experiment, because there was no known way to distinguish between hidden variables and action at a distance. But in his spare time, John Bell came up with a complex setup using a pair of differently aligned detectors, focusing on spin, to produce a set of values that, should they fall outside a defined range known as "Bell's inequality," would demonstrate that there

▼ CERN, the European Organization for Nuclear Research, now home of the Large Hadron Collider, is where John Bell developed his test for entanglement.

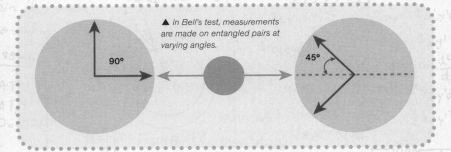

▲ In Bell's test, measurements are made on entangled pairs at varying angles.

were no hidden variables and that quantum physics had withstood Einstein's last assault. This was, at the time, impractical to undertake, but the stage was set to put Einstein on trial.

Bell's intention was never to show that Einstein was wrong. He felt instinctively that Einstein's logic should beat Bohr's less concrete arguments. If anything, he wanted quantum theory as it stood to be discarded. However, with admirable objectivity, his approach did not push the outcome in any particular direction. He had produced clear alternative outcomes. If an experiment could ever be devised to meet his test, then it should decisively prove whether spooky action at a distance existed or not.

▼ Northern Irish physicist John Bell devised the test for entanglement's "spooky connection" in his spare time, while working at CERN.

ASPECT'S AMAZING APPARATUS

John Bell was a theorist — and entanglement had no connection to his day job at CERN. But in the early 1970s, a young French physicist called Alain Aspect decided to take up Bell's challenge. In his free evenings, Aspect mulled over the practicalities of testing Bell's inequality.

Later, Aspect designed an apparatus using the pair of detectors that Bell's experiment required. The most difficult problem was ensuring that these two detectors were not communicating in some other way, making it unnecessary for entanglement to be able to communicate instantly. Aspect decided that he could ensure this didn't happen by changing the orientation of the detectors while the entangled photons were in flight.

This involved flipping the detectors millions of times a second. Aspect did this by using a transducer — similar to the part of a loudspeaker that makes the

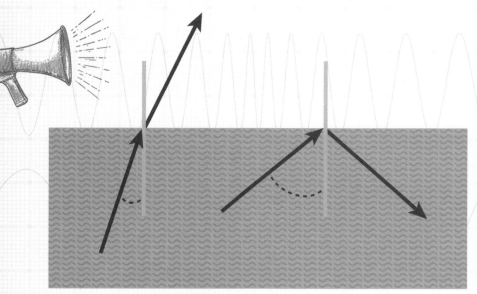

▲ The light passes out of the water when it is not squeezed, but is reflected back when the transducers put the water under pressure.

▲ Alain Aspect

cone vibrate to and fro — to squeeze and relax a container of water repeatedly. This pressure altered the water's refractive index, changing the direction of light traveling through it. Aspect used this to switch the direction of light traveling between the detectors 25 million times a second, leaving no time for information about their settings to travel from one end of the apparatus to the other at the speed of light. The results showed that there were no hidden variables. Entanglement was real. More sophisticated experiments carried out since have always confirmed Aspect's results.

Alain Aspect dreamed up his experiment during the three years he spent as an aid worker in Cameroon after gaining his doctorate in physics.

INSTANT COMMUNICATION

As soon as most people come to understand the implications of quantum entanglement they see a potential application. Because a change in one particle is immediately reflected in the other, however far apart they are, this seems to imply that entanglement could send a message instantly across any distance.

On Earth, message times are rarely an issue in normal communications. But entanglement would be immensely useful in communicating across space.

FTL MESSAGES
• • • • • • •

Remarkably, special relativity shows that if instant messages did exist, there is a mechanism by which we could transport information backward in time. Special relativity says that the time on a moving object — a spaceship, say — runs more slowly than it does on the place it moves with respect to. So, for instance, if a spaceship heads off from Earth at close to the speed of light, it might be noon on Earth but half an hour earlier on the spaceship. An instantaneous message would travel back half an hour in time reaching the ship. The effect is symmetrical, so from the ship's viewpoint, the Earth's clocks run slowly. So the message, sent back from the ship at 11:30, would arrive even earlier on the Earth.

◀ *Earth and Mars: station on Earth sends a message; 20 minutes later the message arrives at Mars; 40 minutes later the response arrives back on Earth.*

▲ As time passes slower on the ship as seen from Earth, and vice versa, an instant message from the ship moves into the past.

Special relativity shows that the passage of time depends on the relative speed of movement — passing an instant message between moving objects can displace the message through time.

Mars missions, for example, cannot control planetary rovers remotely because it can take 20 minutes for a message to get from the Earth to Mars even at the speed of light. It would also be very valuable on a small scale in computing, where even tiny delays in transmitting information slow down processes.

Sadly, although entanglement does involve a kind of instant communication, we have no way of controlling the information sent, which is entirely random. Despite many efforts to find a way to harness the effect for this purpose, no one has been able to get around this limitation.

QUANTUM ENCRYPTION

Entanglement may not make instant communication possible, but it still has the potential for impressive applications, such as in encryption.

It has long been possible to make totally unbreakable encryption by adding a different random element to each character of a message, but such a "one-time pad" can be decoded only by sending the key to the receiver, and there is always a danger that this key will be intercepted. With quantum entanglement, it is possible to provide a random encryption key and also to make it impossible to intercept that key.

▲ *Encryption, such as that used by the Second World War Enigma machines and secure websites, can in theory be broken because there are patterns and mathematical clues, but the encryption provided by a one-time pad is completely random.*

▶ *The encryption used on the internet is extremely secure, as it involves finding the factors of huge numbers, but it is breakable.*

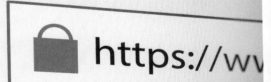

SSL Certificate

https://ww

The value communicated between entangled particles, such as "spin up" or "spin down," is randomly selected, which automatically generates an unbreakable encryption key. The value doesn't exist before it is shared, so there is no preexisting pad to be intercepted. And if a snooper should intercept the flow of entangled particles, forcing them into a value before they are received, the collapse of the entanglement can be detected and the communication stopped before information has been leaked.

The basic unbreakable one-time pad was devised by the American engineer Gilbert Vernam and the future chief of the U.S. Army Signal Corps Joseph Mauborgne in the early part of the 20th century — but without the benefit of quantum key distribution.

ENTANGLED SATELLITES

The use of encryption based on quantum entanglement is close to being commercially available. Making this a reality has proved a significant challenge, because it isn't easy to keep particles entangled. If they have a chance to interact with other particles, they will usually lose their entanglement.

Early experiments involved sending entangled photons across Vienna — one demonstration sent an encrypted message to transfer money to a bank. Over time these experiments have been extended to cover greater and greater distances. The latest development makes it likely we could see a widespread use of this technology: a satellite providing entangled particles.

Such a satellite effectively sends out the raw materials for entanglement-based encryption. Two locations on Earth each receive one of an entangled pair of particles, time after time, in a stream. This gives them the key to communicate securely. In August 2016, China launched the first quantum communications satellite, known as "Mozi," which has been described as the first step toward a "quantum internet," distributing entangled particles to sites as far as 746 miles (1,200 km) apart.

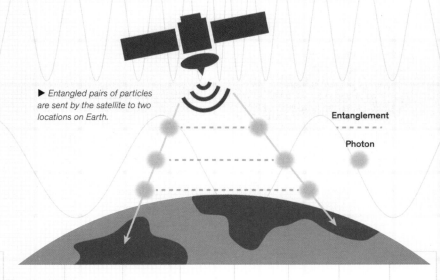

▶ Entangled pairs of particles are sent by the satellite to two locations on Earth.

Entanglement

Photon

For the Vienna bank-transfer demonstration, Austrian physicist Anton Zeilinger and his team ran optical cables through the old sewers that featured in the movie The Third Man.

▲ The Mozi satellite (officially QUESS — Quantum Experiments at Space Scale) launching.

QUANTUM TELEPORTATION

One of the more impressive capabilities of entanglement is that it makes teleportation possible — a small-scale version of a *Star Trek* transporter.

Usually, it isn't possible to find out the exact state of a quantum particle, because it is in a superposition of the different states it can be in, such as spin up and spin down, and observing it fixes its state. However, by using entanglement to teleport the state without ever knowing what it is, it is possible to transfer properties from one quantum particle to another. At the end of the process, the remote particle becomes an identical copy of the original.

'Even if teleportation were possible for people, it wouldn't be attractive. It wouldn't transport you: it makes an exact copy and destroys the original.'

▲ The hypothetical wormhole would provide a mechanism for faster-than-light travel, but teleportation does not make this possible.

TELEPORTING COMPUTERS

Quantum teleportation is unlikely ever to be able to transfer more than a single molecule at a time. However, the process is very important for the functioning of quantum computers (see page 284), which use teleportation to transfer information between parts of the system.

Because entanglement is instantaneous, it has been suggested that quantum teleportation is like traveling faster than light — but the process involves both instantaneous entanglement *and* sending data by conventional means, such as radio or internet, so in practice it doesn't break the light-speed barrier.

▶ Development and research of a quantum computer.

QUANTUM ZENO EFFECT

As we will discover later, quantum effects have been discovered in biological systems. In a sense, all of biology is quantum-based, because it depends on atoms and molecules, but the new discoveries are explicit mechanisms using quantum effects.

One of these is the way that birds can find their way over long distances by following the Earth's magnetic field. It is thought that this may depend on the entanglement of electrons in the birds' eyes (see page 240), though for a long time it seemed highly unlikely that electrons could stay entangled long enough to be useful in a warm, wet biological environment.

However, there is a strange phenomenon called the quantum Zeno effect, named after the Greek philosopher who tried to prove that change was nonexistent. Although a quantum particle is usually a collection of probabilities, its properties take on specific values when it is observed. As a result, repeatedly observing it, before a specific value can change too much, makes it like the watched pot that never boils. It may be that this effect keeps the birds' electrons entangled.

Zeno's most famous argument against motion involves Achilles and the tortoise. The two are racing, and Achilles gives the tortoise, which is much slower, a head start. The hero soon reaches the point that the tortoise had got to when he started, but the tortoise has moved on. Achilles gets to its new point even more quickly... but the tortoise has moved on again. Despite running faster than the tortoise, Achilles can never catch it.

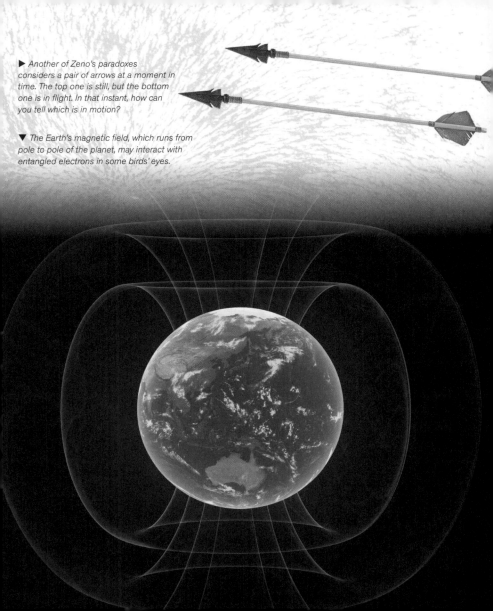

▶ Another of Zeno's paradoxes considers a pair of arrows at a moment in time. The top one is still, but the bottom one is in flight. In that instant, how can you tell which is in motion?

▼ The Earth's magnetic field, which runs from pole to pole of the planet, may interact with entangled electrons in some birds' eyes.

CHAPTER 7
THE GOLD STANDARD

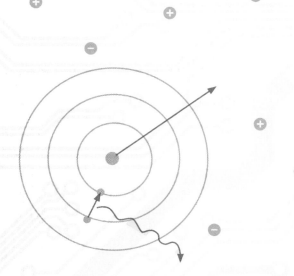

▶ Part of one of the vast detectors in the Large
Hadron Collider, a key experiment in developing
our understanding of particle physics.

ANTIMATTER EVERYWHERE

All fundamental particles are quantum particles — but particle physics is not primarily concerned with the quantum nature of those particles. Instead, it focuses on establishing just what the fundamental particles are, and their relationships to each other, a study that to date has culminated in the standard model of particle physics.

It is essential here to extend Dirac's idea of the positron as the antiparticle of the electron (see page 124) and provide an antiparticle for every "normal" particle. Just as particles make up matter, antiparticles are described as antimatter. Current cosmological theories suggest that matter was originally formed from energy. For this to happen, there should be as much antimatter in the universe as there is matter, as energy converts into matter by producing particles in matter–antimatter pairs.

◀ The early emergence of matter is reflected in the so-called afterglow of the big bang, the cosmic microwave background.

Because we can't see all of the universe (leaving aside the question of whether the universe is infinite or not), there could, in principle, be a separate "antimatter half" — but it isn't obvious how this could have come about.

OH, DEAR, WHERE CAN THE MATTER BE?

There is considerable speculation over where all the antimatter in the universe has gone, because we see only small amounts in nature. The most strongly supported explanation is that there is a small asymmetry in the physics that would produce slightly more matter than antimatter, and that this resulted in the matter we experience.

▼ *Each of the fundamental matter particles has an equivalent antiparticle, as do composite particles such as protons and neutrons, made up of the smaller matter or antimatter quarks. The proton's antiparticle, the antiproton, is negatively charged, while the antineutron has a neutral electrical charge, but differs from the neutron in other quantum properties.*

	Charge	Matter			Antimatter			Charge	
Quarks	$+\frac{2}{3}$	**u** Up quark	**c** Charm quark	**t** Top quark	**ū** Anti-up quark	**c̄** Anti-charm quark	**t̄** Anti-top quark	$-\frac{2}{3}$	**Quarks**
	$-\frac{1}{3}$	**d** Down quark	**s** Strange quark	**b** Bottom quark	**d̄** Anti-down quark	**s̄** Anti-strange quark	**b̄** Anti-bottom quark	$+\frac{1}{3}$	
Leptons	0	νe Electron neutrino	$\nu\mu$ Muon neutrino	$\nu\tau$ Tau neutrino	$\bar{\nu}e$ Anti-electron neutrino	$\bar{\nu}\mu$ Anti-muon neutrino	$\bar{\nu}\tau$ Anti-tau neutrino	0	**Leptons**
	-1	e^- Electron	μ Muon	τ Tau	e^+ Positron	$\bar{\mu}$ Anti-muon	$\bar{\tau}$ Anti-tau	$+1$	

ACCELERATORS AND COLLIDERS

Most experimental particle physics involves an apparently childlike approach to science. We can't see what is going on directly, so we resort to the equivalent of understanding the workings of a clock by hitting it with a sledgehammer and filming all the component parts as they fly out of it in slow motion.

The earliest curved accelerators were cyclotrons, conceived in Germany in the 1920s and first operated by Ernest Lawrence in the U.S. in 1932.

The particle physicists' sledgehammers are accelerators and colliders (usually in combination). These work by accelerating particles up to extremely high speeds, then colliding them — with

▲ *Ernest Lawrence holding his first cyclotron in the 1930s.*

HADRON HAPPENINGS

The "hadron" in the name of the LHC refers to particles made up of two or more of the fundamental matter particles called quarks: these include protons, neutrons and mesons. This is to contrast with leptons, such as electrons, which are fundamental particles in their own right. The LHC probably used "hadron" to distinguish it from a predecessor, which accelerated electrons and positrons in the same tunnel. In practice, the hadrons that are accelerated are protons.

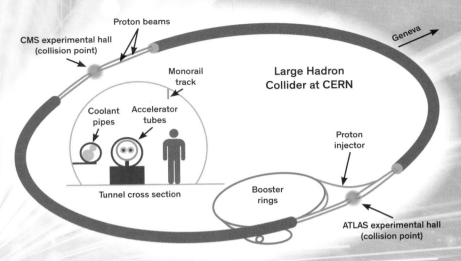

Proton beams

CMS experimental hall
(collision point)

Geneva →

Monorail
track

Large
Hadron
Collider at CERN

Coolant
pipes

Accelerator
tubes

Proton
injector

Tunnel cross section

Booster
rings

ATLAS experimental hall
(collision point)

another moving particle or with a fixed target — and capturing the resultant spray of activity after the collision and analyzing it.

Charged particles are accelerated by passing them through electrical and magnetic fields. Originally this was in a straight line (in linear accelerators), but it was soon realized that particles could be accelerated in a much smaller space if they were repeatedly given a boost as they passed around a curved track. Best known of all the accelerators is the massive Large Hadron Collider (LHC) at CERN, a 17-mile-long (27 km) underground loop that accelerates particles to within a fraction of the speed of light before colliding them in vast detectors.

▲ *Cutaway diagram of the Large Hadron Collider at CERN.*

▼ *Millions of events per second are captured in a high-powered collider.*

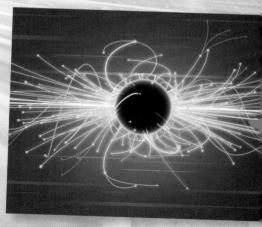

PARTICLES FROM THE COSMOS

Even the latest accelerators can't produce the same acceleration as nature, and the first high-speed particles to be studied were in cosmic rays, high-energy streams of particles that reach the Earth from deep space and collide with atoms in the atmosphere.

Positrons and a range of other quantum particles were first discovered in cosmic rays. The rays have also been used to demonstrate the time-bending effects of special relativity. Because time is slowed for fast-moving objects, particles with short lifetimes survive far longer than they should in the rays. For example, when cosmic rays hit the upper atmosphere they produce particles called muons, an elementary particle similar to a heavier version of an electron. These

▲ *The installation of the Big European Bubble Chamber at CERN, typical of the large bubble chambers used in the 1970s to study particles produced in high-energy accelerators.*

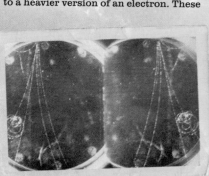

◀ *Early cloud chamber images in the 1930s.*

▲ *Studies of cosmic rays soon left the Earth behind, with special balloons being used to observe them in the upper atmosphere.*

should decay almost immediately, never getting further, but because of this "time dilation" effect, muons can be detected at ground level.

Although cosmic rays have more power than accelerators, scientists prefer to work with the latter, because the particles involved and the energy applied to them are under their control.

MYSTERY RAYS

• • • • • • •

We still aren't sure of the source of all cosmic rays, though it is thought that at least some of them originate in supernovas and speed across the galaxy before hitting the Earth.

The first serious studies of cosmic rays were made in 1909 by German physicist and priest Theodor Wulf on the Eiffel Tower.

Working with the American physicist Seth Neddermeyer at the California Institute of Technology, Carl Anderson discovered muons in 1936. This is the same Carl Anderson who was the first to find a positron (see page 126).

THE PARTICLE ZOO

There was increasing concern as the particles generated in cosmic-ray showers and accelerator collisions were studied.

Around the mid-1930s, there was a clear and simple picture of the fundamental particles that made up the universe: protons, electrons and neutrons in matter, the odd little particle called the neutrino, and photons of light. Yet the showers of particles produced by collisions contained a whole host of

1897	1899	1919	1932	1932
Electron discovered by J.J. Thomson	Alpha particle discovered by E. Rutherford in uranium radiation	Proton discovered by E. Rutherford	Neutron discovered by J. Chadwick	Antielectron (or positron), the first antiparticle, discovered by C.D. Anderson

new particles, so much so that the situation tended to be referred to as the "particle zoo."

The first new particle to be added to the list was the muon, discovered in cosmic rays as early as 1936, while from the 1940s onward they were joined by pions, kaons (both of which were eventually identified as types of meson), J mesons, tau mesons, upsilon mesons and more.

What had once been a universe built from a handful of fundamental particles seemed now to be chaotically underpinned by a whole mess of particles with no clear relationships. This particle zoo was both fascinating and frustrating for those who observed it.

NEW PARTICLES

• • • • • • • •

Muons, as we have seen, are in the same family as electrons, but much heavier. Many more of the new particles were mesons. These are made up of pairs of the fundamental matter particles called quarks — each meson has one quark and one antiquark. The majority of meson types have different "flavors" of quark and antiquark (as we'll see on page 190, quarks come in six different flavors), but where the flavors are the same the meson has only an extremely short lifetime before the particle and antiparticle recombine, annihilating to produce pure energy.

1936	1969	1983	1995	2012
Muon (or mu lepton) discovered by S. Neddermeyer, C.D. Anderson, J.C. Street and E.C. Stevenson	Partons (internal constituents of hadrons) observed in deep inelastic scattering experiments between protons and electrons at SLAC	W and Z bosons discovered by C. Rubbia, S. van der Meer and the CERN UA1 collaboration	Top quark discovered at Fermilab	A particle exhibiting most of the predicted characteristics of the Higgs boson discovered by researchers at CERN's Large Hadron Collider

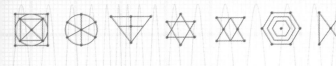

SYMMETRY RULES

A hint of what was going on in the particle zoo came from the mathematical concept of symmetry. Nature often seems to involve symmetry, and it was hoped that, by applying more sophisticated concepts of symmetry than the simple mirror symmetry of matter and antimatter, it would be possible to find an underlying pattern to explain the apparently unstructured mix of different particles in the zoo.

The American physicist Murray Gell-Mann gave his approach the evocative title "the eightfold way," suggested by a term from Buddhism. It was "eightfold" because it seemed possible to organize particles into groups of eight based on their quantum properties, such as spin, corresponding to the different possible arrangements of three-by-three matrices. This reflected the symmetry of a mathematical structure called SU(3) or "special unitary group of degree 3."

The underlying symmetry seemed to suggest that there were groupings

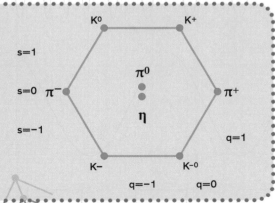

▶ The eightfold way came out of two variants of two quantum properties: the familiar electrical charge (q) and a new property invented to make the system work, called strangeness (s).

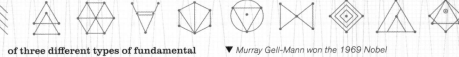

of three different types of fundamental particles, probably including electron-like particles, photons and something that lay behind the rest. This mathematical analysis provided the first real evidence that there might be something more basic than the familiar protons and neutrons, and perhaps enabled the particle zoo to be simplified, because many of the new particles were in fact constructed from combinations of simpler particles.

Symmetry is key to many aspects of physics. The conservation laws, for example, reflect various symmetries in time and space. Conservation laws state that various quantities, such as the amount of energy in a closed system, cannot change. The German mathematician Emmy Noether showed in the early part of the 20th century that each such law results from a symmetry. The conservation of momentum, for example, occurs because physics is symmetrical to translations through space — you can move an experiment to the left or right and it still works. The mathematics to prove this is fiendishly complex, but it is thoroughly proved.

If symmetry were carried to its extreme, we would expect far more consistency in the universe than we actually discover. The complexity of

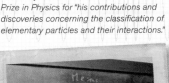

▼ Murray Gell-Mann won the 1969 Nobel Prize in Physics for "his contributions and discoveries concerning the classification of elementary particles and their interactions."

what is observed, from the different forces of nature to the detailed structure of galaxies, is often ascribed to "spontaneous symmetry breaking," where a symmetrical but unstable state decays into an asymmetric, stable state. For example, a pencil balanced on its point is symmetrical as seen from above. But the slightest movement or breeze will cause it to collapse, ending up with the asymmetric form of the pencil lying on its side, pointing in an unpredictable direction.

It is thought that shortly after the big bang, there was a "soup" of free quarks. But once quarks started to join they remained together, as the force that holds them together, the strong nuclear force, has the strange property that it grows as they move farther apart. Quarks typically form either pairs of quarks and antiquarks in mesons, or triplets. Two up quarks and one down quark form a proton, while one up and two down form a neutron.

ACES AND QUARKS

By 1964, Gell-Mann was ready to describe a lower level of fundamental particle that he called "quarks," that combined in trios to make up neutrons and protons, and in pairs to form mesons. It is often said that Gell-Mann took the word from James Joyce's modernist novel *Finnegans Wake*, which contains the phrase "Three quarks for Muster Mark!" In fact, Gell-Mann had already dreamed up the sound "kwork" for the particle before adopting Joyce's spelling.

▼ Illustration of the atomic model showing the structure of a proton or neutron composed of quarks.

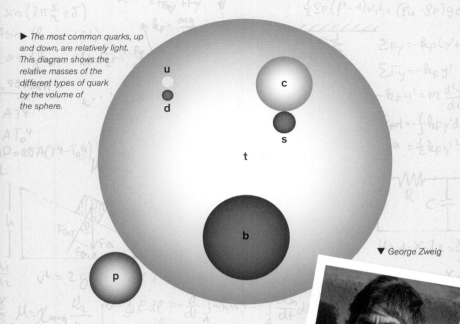

▶ The most common quarks, up and down, are relatively light. This diagram shows the relative masses of the different types of quark by the volume of the sphere.

u
c
d
s
t
b
p

▼ George Zweig

As is often the case, another scientist was following a similar path. A second American physicist, George Zweig, working at CERN, had a similar idea to Gell-Mann's, and published his theory in the same year. He described particles called "aces," giving them this name as he believed that there were four in total.

The theory would be experimentally confirmed within a few years, and a total of six different flavors of quark — up, down, strange, charm, top and bottom — were identified. The last to be found, the very heavy top quark, only turned up in 1995.

There is still no agreement over whether "quark" should be pronounced "kwark," as the spelling suggests, or "kwork," as the man who named it intended.

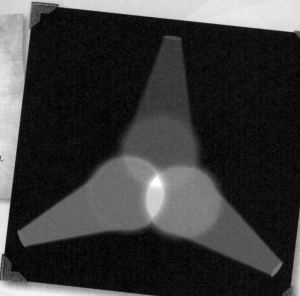

QUANTUM CHROMODYNAMICS

Gell-Mann and his colleagues wanted to extend the approach of QED (see page 130) to cover quarks. But the challenge proved distinctly more complex than that of quantum electrodynamics. This was because all quarks are not identical, but rather come in three different "colors" — red, blue and green (which is why the system is called "quantum *chromo*dynamics"). There is no suggestion that quarks are colored in reality — it is just a label.

The label was, however, well chosen: just as red, blue and green light combine to make white, quarks, when combining to

PARTICLE GLUE

• • • • • • •

Particles called gluons carry the strong nuclear force, the member of the four fundamental forces of nature (see page 196) that binds quarks together (a similar role to photons in QED). There are eight different gluon variants, a number given by the quantum superposition of pairs of color and anticolor states. This means that the family of gluons neatly explains Gell-Mann's eightfold way.

form a particle, are always in a combination of different quark colors that produces white. When, for example, three quarks are joined to make a proton or neutron, there must always be one each of red, blue and green. In the case of mesons, one quark must be the anticolor of the other. (Anticolors are a reminder that these colors aren't real. They are just labels. An anticolor wipes out an original color to make white: there is no equivalent in a paintbox.)

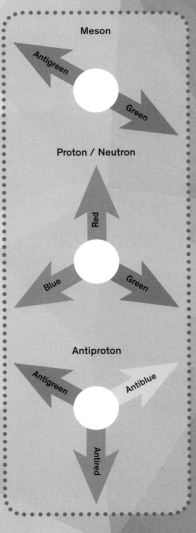

▶ Quarks combine to make particles such as mesons and protons so that their color combinations produce white.

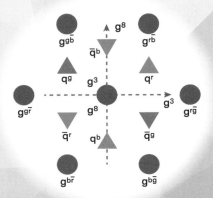

▲ The eightfold-way structure of quarks and gluons. There are three colors of quarks and three of antiquark (anticolors have a bar over the color letter). Six gluon types consist of color mixes, while a further two are special "uncharged" gluons.

MIGHTY MORPHING NEUTRINOS

**The most mysterious of the early fundamental particles was the neutrino.
It was predicted in 1930 to explain the way in which energy seemed to go
missing when some atomic nuclei decayed, but the neutrino proved so difficult
to spot that it was not until 1956 that its existence was confirmed.**

Neutrinos have almost no mass and no
electrical charge, which means that the
billions of them that are emitted by the
Sun and pass through your body every
second leave no trace.

Neutrinos were
an uncomfortable fit
with the gradually
developing picture
of the fundamental
particles, until it
was discovered
that there was
more than one

type, with variants corresponding to
the electron and its heavier cousins the
muon and tau particles, having the same
quantum states such as charge and spin,
but with far higher mass.

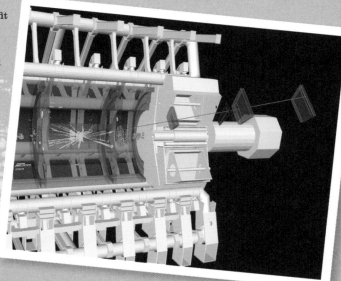

▶ *CERN's neutrino
experiment, called
OPERA, sent the
particles from CERN in
Switzerland 455 miles
(732 km) through the
Earth to Gran Sasso
in Italy.*

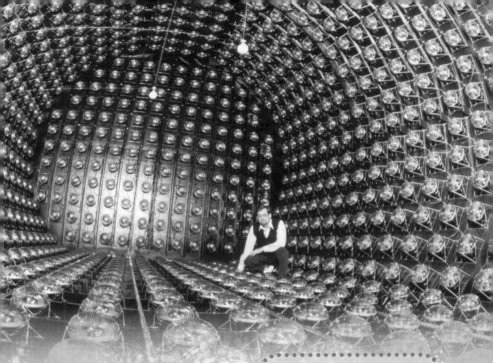

▲ Neutrino telescopes deploy vast arrays of detectors to spot the tiny flashes that occur when a neutrino interacts with a vat of fluid sited far underground to avoid confusion with other particles.

Neutrinos made big news in 2011 when a CERN experiment seemed to show they traveled faster than light, but the result proved to be a side effect of a faulty connection in the equipment.

MORPHING MEANS MASS

• • • • • • • • • • • • • • •

Until recently it was assumed that neutrinos had no mass, but there was growing evidence that they undergo a strange modification in flight, switching between the different possible types (this was experimentally confirmed in 2013). The physics behind this change requires neutrinos to have mass, although this remains extremely small.

THE STANDARD MODEL

The combination of Gell-Mann's eightfold-way theory and discoveries from particle experiments resulted in a four-by-four matrix of fundamental particles that appear to have no small constituents. Each has an antiparticle (though some particles, such as photons, are their own antiparticle).

Six quarks combine with six leptons as particles involved in matter, though of the leptons it is only the electron, in practice, that has a significant role in forming all the matter in everyday life. Four particles

◄ Satyendra Nath Bose.

The word "boson" is often confused with the nautical term "bosun" — but the particle type is named after Indian physicist Satyendra Nath Bose, not a sailor.

THE FOUR FORCES

• • • • • • • •

The particles of the standard model combine with four fundamental forces to provide the backbone of physics. Three of the forces are available to quantum physics. Most familiar is electromagnetism. Carried by the photon, this provides all the familiar interactions of light and matter, from our ability to sit on a chair to our ability to see. The strong nuclear force, carried by gluons, links quarks together and also allows protons to stay together in the nucleus. The weak nuclear force, carried by W and Z bosons, is involved in nuclear reactions and neutrino reactions. The fourth force is gravity, but this is incompatible with quantum theory and currently has to be treated separately. Our best description of the working of gravity is given by the general theory of relativity.

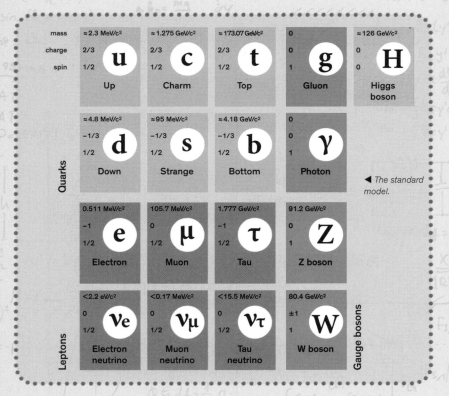

The standard model.

mass	≈2.3 MeV/c²	≈1.275 GeV/c²	≈173.07 GeV/c²	0	≈126 GeV/c²
charge	2/3	2/3	2/3	0	0
spin	1/2	1/2	1/2	1	0
	u Up	**c** Charm	**t** Top	**g** Gluon	**H** Higgs boson
	≈4.8 MeV/c²	≈95 MeV/c²	≈4.18 GeV/c²	0	
	−1/3	−1/3	−1/3	0	
	1/2	1/2	1/2	1	
	d Down	**s** Strange	**b** Bottom	**γ** Photon	
	0.511 MeV/c²	105.7 MeV/c²	1.777 GeV/c²	91.2 GeV/c²	
	−1	0	−1	0	
	1/2	1/2	1/2	1	
	e Electron	**μ** Muon	**τ** Tau	**Z** Z boson	
	<2.2 eV/c²	<0.17 MeV/c²	<15.5 MeV/c²	80.4 GeV/c²	
	0	0	0	±1	
	1/2	1/2	1/2	1	
	νe Electron neutrino	**νμ** Muon neutrino	**ντ** Tau neutrino	**W** W boson	

Quarks · Leptons · Gauge bosons

known as gauge bosons, which interact with their own kind far less than matter particles, carry the various forces with which these particles interact.

A 13th particle is now often added to the model — the Higgs boson (see page 200). As yet, the standard model is our best picture of the underlying building blocks of nature. There are some problems — it does not cover dark matter (see page 206), or work with gravity, which may require far more particles (see page 296) — but the standard model is among physicists' greatest achievements.

WHY DO PARTICLES HAVE MASS?

It may seem silly to ask why particles have mass. It is surely just one of their properties — why shouldn't they? However, mathematics has increasingly come to drive our understanding of physics, and the math worked better if particles didn't have mass. The only way to match reality while sticking with the math was to add something else.

▲ The first hints of a Z boson emerged in CERN's Gargamelle bubble chamber, now on display outside the laboratory complex.

In quantitative terms, the main "something else" was obvious. It was energy. Einstein's $E=mc^2$ had demonstrated the interchangeability of mass and energy. Even if the quarks in a proton or neutron were massless, the resultant particles would still have pretty much the same mass as they have in practice. This is because there is so much energy tied up in the gluons' efforts to hold the quarks together that it nearly corresponds to the measured mass of the proton and neutron.

Other particles, though, such as the electron, and the Z and W bosons that carry the weak nuclear force, had no such excuse for having mass. If the standard model were to survive, there had to be something else out there, giving mass to particles.

> The weak nuclear force is unique in having three different bosons as carriers: Z, W + and W -.

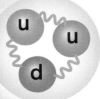

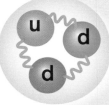

▲ The proton consists of two up quarks and one down quark, linked by gluons. The neutron consists of one up quark and two down quarks, linked by gluons.

▼ The Z and W bosons were identified in CERN's Super Proton Synchrotron, originally an accelerator in its own right and now a kick starter for particles entering the LHC.

THE HIGGS FIELD

Given that most of physics was now described by fields, such as the electro magnetic field, an easy patch to the standard model was to add yet another field that provided a kind of drag on various particles, giving them the mass that they needed. This concept was developed by a total of six physicists: Robert Brout, François Englert, Peter Higgs, Gerald Guralnik, Richard Hagen and Tom Kibble. Strictly speaking, the new field should be the Brout Englert

▲ The huge Atlas and Compact Muon Solenoid (CMS) detectors on the LHC were crucial to the discovery of the Higgs boson.

Higgs Guralnik Hagen Kibble field. But thankfully (for the rest of us if not for the five omitted physicists) it has become known as the "Higgs field."

The concept was devised in 1964 and gradually gained in theoretical standing as the standard model was developed, though it would not be experimentally supported until 2012, when the Higgs boson was discovered.

▲ François Englert

▶ The Higgs field acts in a similar fashion to friction to give particles apparent mass.

It's often said that the Higgs boson gives particles their mass. It doesn't — that's the job of the field. The boson is just a ripple in the field that has no consequence for the rest of the standard model — a phenomenon you would expect to exist were the field there — so it is a way to support the field's existence.

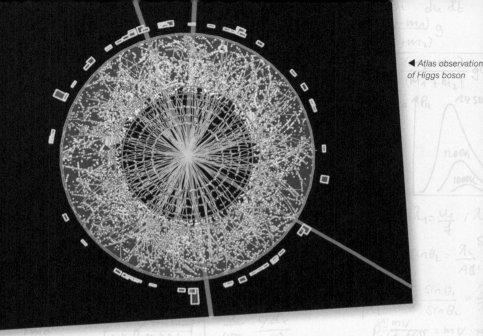

LHC BREAKTHROUGH

The discovery of the Higgs boson in the Large Hadron Collider at CERN — which was the main reason why the accelerator was built — became a worldwide media event, even though the media struggled to understand and explain what had been discovered.

The experiments took place in July 2012, but the sheer volume of data involved meant that the announcement of the result could not be made until March 2013. The result was a particle with a mass appropriate for a Higgs boson (the exact expected mass cannot

be calculated, just a range in which it might fall). It was almost certain that the result was not a random glitch, though it's impossible to prove that it was indeed a Higgs boson.

The Higgs boson was expected because other quantum fields had traveling peaks of energy known as "excitations" that we describe as particles — so, for example, a photon is an excitation in the electromagnetic field. Although there is no way to observe the Higgs field, the presence of an apparent Higgs boson increases the chances that it exists.

The discovery gave CERN another entry to add to its impressive collection of Guinness World Records for the rather specific achievement of "the first proof of the existence of a Higgs boson." Other records that CERN holds include: the largest scientific instrument (the Large Hadron Collider is just under 17 miles/ 27 km in circumference); the world's most powerful particle accelerator; and the highest man-made temperature of 5 trillion K.

The Higgs discovery was described as "5 sigma." Many interpreted this statistical term as meaning there was only a 3-in-10-million chance this wasn't the Higgs boson. In fact it indicated a 3-in-10-million chance of the result occurring randomly — a different proposition.

▼ Simulated proton collision experiment at CERN

SUPERSYMMETRY

Although the standard model has been very successful, it is still clumsy, providing no relationship between matter and force-carrier particles, and it is highly dependent on plugging in values from observation rather than the underlying theory. It is entirely possible that this reflects a fundamental flaw, but equally it could be that the model doesn't go far enough.

The most popular attempt to shore up the standard model is called "super-symmetry." As the name suggests, this relies even more on the application of the mathematics of symmetry (see page 188) and would provide a bridge between the different families of particles. However, this apparent simplification would paradoxically be at the cost of a major increase in complexity, as each particle would have a supersymmetric partner from the other family — matter particles

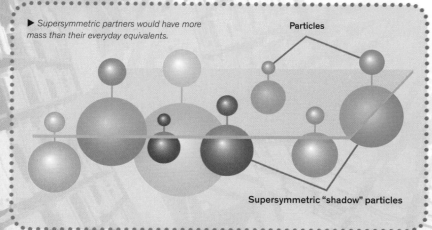

▶ Supersymmetric partners would have more mass than their everyday equivalents.

Particles

Supersymmetric "shadow" particles

would have force-carrier partners, and force carriers would have matter-particle equivalents.

Supersymmetry, also known familiarly as "SUSY," predicts that the partner particles should have both larger mass and a different spin from existing particles. Fermions, such as quarks and electrons, have quantum spins with a multiple of 1/2, while bosons, such as photons, have integer spins. (Confusingly, spin 1/2 means the particle has to "rotate" twice to get back to the same state, while spin 1 means a single "rotation" achieves this.) The supersymmetric particles have spins that are 1/2 lower than their normal equivalents — so selectrons, for example, are spin 0.

While the theory has mathematical neatness, and appeals to those who drive their physics by numbers alone, no evidence has yet been uncovered that these particles exist.

Supersymmetry has strong links with string theory (see page 304).

MEET THE SQUARKS

Supersymmetry doubles the number of particles. So, for example, particles such as quarks and electrons would have force-carrier equivalents called "squarks" and "selectrons," while photons and gluons would have matter-like equivalents called "photinos" and "gluinos."

DARK MATTER

Another particle that is missing from the standard model is the particle behind dark matter. Dark matter was suggested in the 1930s by Swiss astronomer Fritz Zwicky to explain the behavior of some galaxies.

All galaxies spin and, like clay on a potter's wheel, you would expect them to fly apart if they spun too fast. Indeed, galaxies seem to be rotating too quickly for the gravitational attraction of the amount of material detected in them to hold them together. Zwicky's idea was to add an extra type of matter, dark matter, that was not visible but had mass and so added mass to galaxies, making them less likely to fly apart. He was largely ignored. But in the 1970s, the work of American astronomer Vera Rubin showed that this was a real issue when she provided a wide range of data showing that galaxies rotated faster than should be possible, given the apparent matter within them.

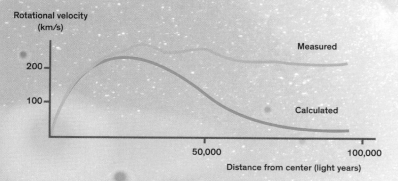

▲ *Theory predicted that the rotational speed of material in a galaxy would quickly tail off farther from the center, but this did not happen, suggesting the outer material should be thrown off into space.*

▲ As a galaxy rotates, centrifugal force pushes components outward, like the chairs on a fairground ride. In the galaxy, it is gravity that takes the place of the chains, holding stars in place.

▼ According to general relativity, mass bends light, acting like a lens. In this image, distant galaxies are magnified by a lens of dark matter, which is mapped out in the photo as an outer blue ring.

Since then, other observational data, such as the way that galaxies bend light due to the mass they contain, has also suggested that there is more to them than (literally) meets the eye. This effect is so large that there appears to be about five times more of this undetectable "dark matter" than ordinary matter in the universe. If that's the case, it is clear that we are some distance from having an effective model covering all of particle physics.

WHY ONE PARTICLE?

A number of candidates have been suggested for dark matter, though the particles have proved very difficult to pin down, as dark matter is thought to interact with ordinary matter only via gravity. Although we think of gravity as powerful, it is, in fact, vastly weaker than the other forces. (Think of picking up a piece of metal with a magnet. The magnet's electromagnetism beats the whole of Earth's gravity.) This makes it harder to detect particles that only interact via gravity than to detect the

▲ Any possible dark matter universe is currently purely speculative.

more familiar particles that interact through electromagnetism.

The most popular suggestion for the nature of dark matter is particles known as "WIMPs" (weakly interacting massive particles). These could in principle be existing particles such as neutrinos (which are almost as hard to detect), or particles called "axions," that are, as yet, purely theoretical. But it now seems unlikely that either of these particles would behave exactly as dark matter appears to do.

Until recently, neutrinos would not have been candidates because they were thought to have no mass, but they were discovered to have a very small mass. Each of the three types of neutrino has a different mass, but combined they add up to only around 1 millionth of the mass of an electron. However, there are vast quantities of neutrinos out there. Every star pumps out many trillions of neutrinos each second.

This make neutrinos seem a likely candidate. However, there are still some problems. One is that even with that vast quantity of neutrinos around, they don't seem to have anywhere near enough mass to account for the effects of dark matter. Another issue is that they are simply too fast. Neutrinos travel at close to the speed of light. But for dark matter to do its job and pull together all the

Dark matter planets make for great science fiction, but the apparent distribution of dark matter makes such small-scale structures (in galactic terms) unlikely.

normal matter to form galaxies, it has to be extremely sluggish in its movement. Astronomers often refer to "cold dark matter," by which they mean moving very slowly. This reflects the way that temperature is a measure of the energy of movement and excitation of particles, so a low temperature is the equivalent of slow motion. Finally, although neutrinos are difficult to detect, we can now do so. Dark matter, however, has remained stubbornly impossible to find.

One possible way to get around these problems is to go supersymmetric. The neutrino's supersymmetric partner, the neutralino, is a better option in principle as it is more massive, but we have no evidence that it even exists.

Considering that there appears to be five times more dark matter than normal matter in the universe, it seems reasonable that the dark matter component could be just as complex as normal matter. Bearing in mind our standard model grid (see page 197), why shouldn't there be a whole range of dark matter particles? It has even been suggested that there may be dark matter planets, illuminated by dark light from dark matter suns — in effect, a parallel dark matter universe — though there is no evidence for this as yet.

A DEAD END?

It was taken for granted that dark matter would be detected — but at the time of writing, no experiment has shown any signs of it. Indeed, the physics community is beginning to wonder if dark matter exists at all.

Clearly something is causing the odd behavior of cosmological structures such as galaxies, but a different kind of matter is not the only possible solution. One early contender was MOND (Modified Newtonian Dynamics). This was based on the idea that we can't necessarily apply the exact behavior predicted by Newton's laws of motion to something the size of a galaxy. We have just tended to assume that the same rules apply on a totally different scale. Only a tiny change would be required to provide some of the effects attributed to dark matter.

More recently it has been suggested that even this is unnecessary. To predict how a body like a galaxy should behave, mathematical approximations have to be made, because the gravitational inter-action of multiple bodies is so complex that we can't predict the outcome perfectly. It has been suggested that the effect assumed to have been caused by dark matter could be down to inaccuracy in the way that the interaction of the material in a galaxy is calculated.

The gravitational interaction of even three bodies is chaotic enough to make complete results incalculable, let alone that of the billions of stars and smaller bodies in a galaxy.

CHAPTER 8

QUANTUM SURPRISES

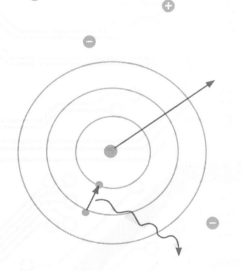

▶ When temperatures approach absolute zero, quantum effects become obvious in full-sized objects.

NATURE ABHORS A VOID

An essential tenet of early physics held that "nature abhors a vacuum." This vacuum, more accurately translated from the Greek as a void, is an absence of anything. This was an article of faith partly because it was felt that any true void would be filled by surrounding matter, and (for some) because it could be used to discredit the atomists, as atomic theory required a void between the atoms.

In a sense, quantum theory brings back this long-discarded idea, thanks to the uncertainty principle (see page 100).

Energy and time make up one of the pairs of properties that are linked by this principle. This implies that over a very small timescale, energy levels can fluctuate immensely — enough for the available energy to produce an electron/positron pair, which will then immediately annihilate.

If empty space is teeming with such particles, all of which exist for an extremely short time, it is impossible to observe them directly. However, we can observe their impact. A phenomenon called the "Casimir effect" means that two flat plates placed

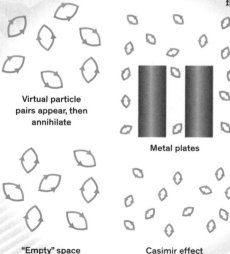

Virtual particle
pairs appear, then
annihilate

Metal plates

"Empty" space

Casimir effect

◀ *The Casimir effect creates an extremely weak force between two close plates as a result of quantum fluctuations.*

very close together — just nanometers (1 nanometer = 1 billionth of a meter) apart — are attracted toward each other. This is because they receive more impacts from the numerous "virtual particles" on the outside than they do from the limited number within the narrow gap. The constant, brief creation of virtual particles means that a totally empty void is unlikely to exist for any measurable period of time.

Usually the virtual particles are not accessible, as the particles recombine before they can be observed. However, if

▲ According to quantum theory, apparently empty space contains a chaotic collection of virtual particles that pop into existence and then annihilate.

close to a black hole, Stephen Hawking suggested that one half of a pair could be pulled into the hole while the other half escapes, producing so-called Hawking radiation.

ARISTOTLE STEALS A MARCH ON NEWTON

Aristotle argued that there could not be such a thing as a void, because, if there were, something would move forever unless it was made to stop. He used this as an example of something he thought impossible, but in fact it's very close to Newton's first law of motion, which includes the statement that an object that is moving will continue to do so unless a force acts on it.

ZERO-POINT ENERGY

We would expect a void to contain no energy — it is, after all, empty. Yet the short-term fluctuations predicted by the uncertainty principle and observed in the Casimir effect mean that "empty" space can never be totally devoid of energy.

The average level of quantum energy in a void is called "zero-point energy." But predicting a value for this level rapidly becomes fraught with infinities and there is no opportunity here to renormalize safely, as was done with QED (see page 140), so estimates for the amount of zero-point energy vary widely.

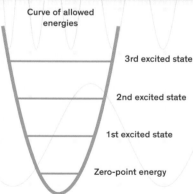

Curve of allowed energies

3rd excited state

2nd excited state

1st excited state

Zero-point energy

▲ *When a quantum object is at its lowest possible energy level, the energy still isn't zero — but to use this energy would require it to go lower still.*

We regularly see announcements that someone has devised a way to harness zero-point energy as a limitless source of power. But the practicalities are daunting. To make use of energy, you need something that has less energy. For example, to harness gravitational potential energy, you need somewhere lower than your starting point. If you have somewhere lower to send water, you can use it to propel a waterwheel or turbine — but on a totally flat plane you could not make use of water power; the energy would be inaccessible. However, by definition, there isn't anything lower than zero-point energy. The only possibility is to make use of the fluctuations — but all attempts to do so have used more energy than can be harvested.

The impossibility of directly accessing zero-point energy (sometimes known as vacuum energy or ground state energy) has not stopped speculation on alternative ways of utilizing its existence. One of the better-supported concepts is the so-called quantum vacuum thruster. Conventional ion thrusters have to carry reaction material to propel out of the engine; a quantum vacuum thruster would use virtual particles as its reaction material so would not be burdened with the extra mass. Many theorists doubt the practicality, but a number of test devices have been proposed.

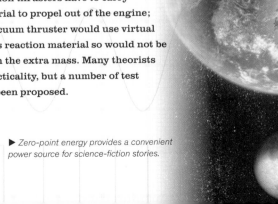

▶ Zero-point energy provides a convenient power source for science-fiction stories.

ABSOLUTE ZERO

Temperature is a measure of the energy in the atoms within a material. This energy is a combination of the kinetic energy of the atoms (their energy of movement) and the energy levels within each atom (the energy state that the electrons are in, which dictates whether they have the potential to make quantum leaps to a different orbit).

You could imagine a state in which every atom was stationary and all the electrons were at the lowest level. Once this was reached there would be no further to go. Such a body would be at the absolute limit of coldness — absolute zero. Theoretically, this would occur at $-273.15°$ Celsius, which was redefined as 0K on the Kelvin scale (which uses the same size units as Celsius).

A QUANTUM LIMIT

Before quantum theory, there seemed no particular reason why this temperature could not be achieved. However, if atoms were all at the lowest energy level and not moving, the uncertainty principle would be breached — you could know both the location (fixed) and the momentum (0) of an atom. As a result, absolute zero can never be achieved.

▼ *Key points on the three most common temperature scales.*

		Absolute zero	Freezing water	Boiling water
Kelvin	K	0	273.15	373.15
Celsius	°C	−273.15	0	100
Fahrenheit	°F	−459.67	32	212

The Laws of Thermodynamics

0. Two objects are in equilibrium (balanced) as far as heat is concerned, if heat can flow from one to the other, but it doesn't.

1. The energy in a system changes to match the work it does on the outside or that is done on it, and the heat given out or absorbed.

2. In a closed system, heat moves from a hotter part of a system to a cooler part — entropy (disorder) stays the same or increases.

3. The entropy of a system at absolute zero is zero. It is impossible to reach absolute zero in a finite set of operations.

The concept of absolute zero was originally based on noting that the measure in a thermometer gets smaller as temperatures fall — at some point it could not get any smaller.

SUPERCONDUCTORS

In 1911, the Dutch physicist Heike Kamerlingh Onnes was experimenting with extremely low temperatures. He was the world's leading expert in low-temperature physics and was curious to see how electrical conductivity would be affected by a reduction in the energy levels of atoms.

Some thought that as the "gas" of electrons carrying electricity lost energy, the resistance of materials would shoot off to infinity. Others, including Kamerlingh Onnes, thought there would be a gradual decline in resistance.

In his 1911 experiment, Kamerlingh Onnes lowered the temperature of mercury down to 1.5K — just above absolute zero — and what he observed shocked the physics world. Neither prediction proved

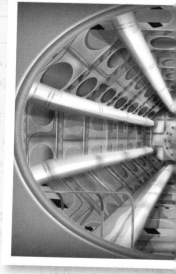

▲ The huge magnets in the Large Hadron Collider rely on superconductors to carry immense currents.

▼ Superconductors have a number of unusual capabilities, one of which — excluding a magnetic field — forces a magnet to float above them, in the Meissner effect.

correct. At 4.2K, the mercury's resistance suddenly and entirely disappeared. Kamerlingh Onnes had discovered the world's first superconductor. A total lack of resistance was hard to prove, but in later experiments a flow of electricity in superconducting material would run for 18 months with no drop in current.

Although superconductors have their limits, because they tend to heat and lose superconductivity when a current is put across them, they facilitate the production of far stronger electromagnets than is otherwise achievable.

▶ Heike Kamerlingh Onnes was awarded the 1913 Nobel Prize in Physics for "his investigations on the properties of matter at low temperatures which led, inter alia, to the production of liquid helium."

▼ Superconducting temperatures for a range of materials.

Superconductor	K	°C
Aluminum	1.2	−271.95
Diamond	11.4	−261.75
Lead	7.19	−265.96
Mercury	4.15	−269
MgB$_2$	39	−234.15
NbN	16	−257.15
Tin	3.72	−269.43
Titanium	0.39	−272.76

Even by the standards of the early 20th century, Kamerlingh Onnes was said to be paternalistic and overbearing with his staff.

ROOM-TEMPERATURE DREAM

Superconducting mercury is impressive, but the ultimate dream would be to have superconductors that operate at room temperature. This would mean, for example, that power cables could carry electricity without loss, or higher currents over greater distances.

Room-temperature superconductors have yet to be achieved, but the highest temperature at which superconductivity is possible has been pushed up several

▲ The aim is to find superconductors that won't need constant cooling with fresh supplies of liquid helium or nitrogen.

As yet there is no good theory to explain the existence of modern high-temperature superconductors.

▼ New ceramic materials provide superconductivity at higher temperatures than ever before.

Compound	K	°C
MgB_2	39	−234.15
$Tl_2Ba_2CuO_6$	80	−193.15
$TlBa_2Ca_3Cu_4O_{11}$	122	−151.15
$HgBa_2Ca_2Cu_3O_8$	128	−145.15
high-pressure H_2S	203	−70.15

times. The initial climb was gradual, to around 30K (-243°C), but in the mid-1980s new ceramic materials, combining barium, copper, yttrium and oxygen, were discovered, allowing a massive leap up to 90K (−183°C). Since then, various other combinations have resulted in superconductors operating at up to 125K — around the −150°C mark. These have the huge advantage that superconductivity can be achieved using commonplace liquid nitrogen rather than the expensive liquid helium that is required at lower temperatures.

▲ At the time of writing, superconducting cables are complex and need constant cooling.

SUPERFLUIDS

When Heike Kamerlingh Onnes cooled liquid helium down to 1.5K in his conduction experiment, he noticed something odd. In the cooling process, the liquid helium clearly bubbled as parts of it exceeded its boiling point of 4.2K, then condensed back. But at 2.17K, the bubbling suddenly stopped.

Kamerlingh Onnes didn't realize it, but the surface of the helium had become a superfluid. Just as extreme cold changes the way electrons travel through a conductor, so it has a quantum effect on the atoms of a liquid, producing a substance with no viscosity — no resistance to movement. Because the atoms in the substance can move totally freely, they do not lose any thermal energy as they move around, so they transmit thermal energy perfectly: they become ideal conductors of heat.

This means that superfluids have seemingly magical properties. Start a ring of superfluid rotating and it will continue to do so as long as it remains in the superfluid state. Even more strangely, when contained in an open vessel, a superfluid will attempt to climb out. A

◀ *An early superfluid escaping from its container.*

▼ By modern standards, Heike Kamerlingh Onnes's laboratory was very basic, but he achieved remarkable results.

Superfluids have few obvious applications, but there have been recent ideas to use them in specialist cooling devices, and as "quantum solvents" that make other substances consisting of small particles clump together and become more visible.

combination of natural movement of atoms, no viscosity, and a quantum effect that sees atoms link together as if they were a single larger entity makes this motion possible. With the correct shape of vessel and a sudden application of heat, the effect is so strong that the superfluid shoots out of the top of the container, producing a self-sustaining fountain.

BOSE–EINSTEIN CONDENSATES

Both superconductors and superfluids are the result of changes in the behavior of collections of quantum particles at low temperatures. A related phenomenon involves the production of a whole new state of matter.

We are familiar with solids, liquids and gases, and most of us will also have heard of the fourth state of matter, plasma, which is produced by heating a gas until the atoms lose electrons and become charged "ions." Because stars are primarily plasma, and make up the bulk of the mass of a galaxy (excluding dark matter), plasma is the most common form of matter in the universe. We also come across plasma in flames and some types of flat-screen TV. But there is also a fifth state of matter at the cold end of the temperature range, known as a "Bose–Einstein condensate."

The standard model (see page 196) divides particles into two groups: fermions, such as electrons and quarks, which obey exclusion principles (see page 120); and bosons, such as photons, which do not. Bosons don't usually interact with each other and many of them can coexist in the same quantum state. In a Bose–Einstein condensate, vast

IONS

••••

Atoms are usually electrically neutral, with the same numbers of electrons as protons. However, many are capable of either adding extra electrons or losing one or more electrons. When this happens, the charged version of the atom is called an ion. These often form in chemical reactions, such as when sodium chloride (salt) dissolves in water to produce sodium and chlorine ions, or when a gas is heated.

numbers of particles are linked together in a way that enables them to act as if they were a collective boson, endowing the material with unusual properties, such as the ability to trap light.

The term Bose–Einstein refers to the model used to describe the collective behavior of these particles — Bose–Einstein statistics, so named as it was

The name comes from Bose–Einstein statistics, one of two ways a group of quantum particles can behave, with the other being Fermi–Dirac statistics.

developed by Satyendra Bose and extended by Albert Einstein. This behavior only applies to particles that don't obey the Pauli exclusion principle. The alternative Fermi–Dirac statistics (derived separately by the two physicists) applies to collections of particles that obey the exclusion principle.

▶ Fermions (shown in green) obey Fermi–Dirac statistics, while bosons (shown in red) obey Bose–Einstein statistics.

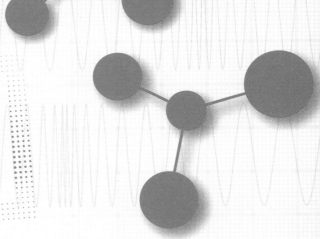

	Quarks			Gauge bosons	
u Up	**c** Charm	**t** Top	**g** Gluon	**H** Higgs boson
d Down	**s** Strange	**b** Bottom	**γ** Photon	
e Electron	**μ** Muon	**τ** Tau	**Z** Z boson	
νe Electron neutrino	**νμ** Muon neutrino	**ντ** Tau neutrino	**W** W boson	

Leptons

SLOW LIGHT

A fundamental aspect of the special theory of relativity is that the speed of light is a constant within any particular material (the ultimate limit of 299,792,458 meters per second — about 983,571,056 feet per second — is light's velocity in a vacuum), but Bose–Einstein condensates have an unusual effect on that speed. Although these have yet to find practical applications, they have proved capable of a remarkable feat: slowing light to walking pace.

Danish physicist Lene Vestergaard Hau, working at Harvard University, made a condensate of sodium ions by cooling them to near absolute zero. A route through

▼ Slow-light experiments require precision laser optical environments.

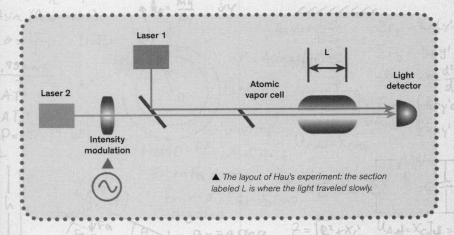

▲ The layout of Hau's experiment: the section labeled L is where the light traveled slowly.

to light, was opened up using a laser, described as a coupling laser, which acts as a kind of optical ladder for a second laser to pull its way through. A second beam then sent through the material was massively slowed — first to 17 meters (about 55.8 ft) per second and then to 1 meter (about 3.3 ft) per second.

An alternative approach to bringing light to a standstill would be to generate a spinning vortex in a Bose–Einstein condensate. If this tiny whirlpool could be made to spin fast enough, the effect would be to produce a kind of optical black hole into which light is dragged and cannot escape, as the fluid is moving faster than the speed of light within it.

THE STOP LIGHT

• • • • • • • •

In a later version of Hau's experiment, the coupling laser was gradually reduced to nothing, leaving the main beam trapped inside the condensate. Restarting the coupling laser enabled the trapped photons to escape. In effect, the condensate produces an entangled mix of light and matter, known as a "dark state."

Hau's experiment had to be abandoned for several days when a film crew used smoke to make the lasers in her laboratory visible on camera.

THE NON-LIGHTSABER

Another example of using low temperatures to produce unusual quantum effects occurred in 2013, leading to a headline proclaiming that "MIT, Harvard Scientists Accidentally Create Real-Life Lightsaber." It sounds impressive, but the headline, inspired by a university press release, was more than a little misleading. Would-be Jedis were going to be disappointed.

Once again, a Bose–Einstein condensate had been employed to make light behave unusually. Photons of light generally ignore each other completely, which is just as well. If all the different streams of photons crossing a room were to collide, the result would be optical mayhem. There's not just the visible light our eyes detect, but also the electromagnetic spectrum. So all the different radio, TV, phone and Wi-Fi signals, each with its own stream of photons, would contribute to the chaos. But in the "lightsaber" experiment, two photons effectively had to stick together to form a kind of light molecule.

As the two photons passed through the condensate, one acted like the coupling laser, which made the second photon interact strongly with its environment and tied the two photons together. Such a pair of photons act as if they have mass and are attracted to each other as long as they remain in the Bose–Einstein condensate. As soon as they escape, the photons return to their normal behavior, though they are frequently entangled. It is hoped that this kind of effect will be useful in photonics, which emulates the capabilities of electronics but uses photons rather than electrons, perhaps leading to much smaller and faster devices. It's interesting physics... but nothing to get Luke Skywalker excited.

Another headline read: "Scientists Finally Invent Real Working Lightsabers" — as if physicists had taken far too long to achieve this.

◀ Lightsabers make great science fiction, but they cannot be made using Bose–Einstein condensates.

MACRO AND MICRO WORLDS

Arguably, the most remarkable quantum experiment is that of our everyday existence in the universe — the macro world. We are surrounded by objects constructed from quantum particles, with each particle exhibiting the weird behavior we have explored — the micro world.

At the same time, quantum particles provide the absolute foundations of reality. Yet the macro world we live in is occupied by ordinary objects that seem to have an unwavering reality. The only uncertainty about, say, a piece of cheese in the fridge is whether or not someone has eaten it. The cheese will not, of its own accord, find itself elsewhere, even though the individual particles that make it up are capable of quantum tunneling (see page 114).

This distinction between the macro and the micro is part of the reason why so many people — Einstein included — have had difficulty with quantum physics. Quantum weirdness undoubtedly happens, but it feels as if it shouldn't. We can say that in larger structures, the interaction with the other quantum particles around them "tames" the quantum oddness, but it still doesn't feel quite right.

The macro/micro distinction is why oddities such as superconductivity are so interesting. These are quantum behaviors observable in macro objects — a mercury wire or a puddle of liquid helium, for example — that we can directly experience.

Despite the apparent conflict, practically nothing we experience involving macro objects would work the way it does without quantum effects. We would have no Sun and stars without quantum tunneling. Matter as we know it would be neither visible nor able to form solid objects without quantum interactions.

◀▼ The macro world we experience behaves as we expect, despite being made up entirely of quantum particles.

◀▼ The micro world of quantum particles is driven by probability and seems to run counter to experience.

QUANTUM ENZYMES

One of the fields in which we are increasingly realizing the importance of quantum effects is biology. A surprising number of biological processes depend on quantum physics, not just in the sense that they involve the quantum particles that make up matter, but because quantum effects directly influence biological outcomes.

An early discovery of this from the 1970s involved enzymes, which are essential, for example, in the process of digesting food. In some cases, it is well established that the catalyzing role of the enzyme involves enabling protons or electrons to tunnel through a barrier, encouraging a reaction to take place. This is necessary,

▶ Biological detergent makes use of the quantum effects of enzymes to break down large molecules such as fats and carbohydrates, making stains on materials easier to remove.

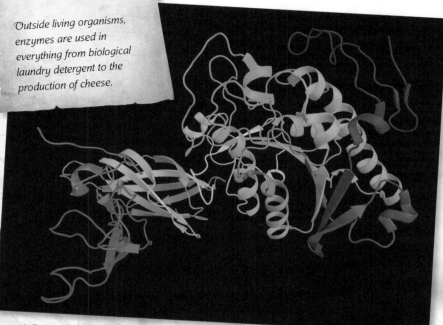

Outside living organisms, enzymes are used in everything from biological laundry detergent to the production of cheese.

▲ Enzymes are complex biological molecules, forming intricate folded shapes.

for instance, when the gut absorbs energy from food. The process could still occur without the help of a quantum effect, because some particles would have enough energy to surmount the barrier, but quantum intervention speeds up the reaction — in many cases making it thousands of times faster.

As well as in the gut, enzymes are active in many reactions that take place in all living things, such as the transmission of chemical signals around an organism or the regulation of the flow of energy-carrying substances.

QUANTUM SURVIVAL

Without these go-faster effects, many organisms, including humans, could not survive. So, just as quantum effects make it possible for us to live by enabling the Sun to function, they also drive our biological existence.

TUNNELING DNA

The molecule DNA (deoxyribonucleic acid) is at the heart of life, storing large quantities of data that provide the genetic instructions for the way that living things grow and reproduce.

This data can be easily duplicated when the spiral staircase structure of the molecule unzips in mid-"tread" and each half is reconstructed as a complete version. This duplication mechanism is one of the points at which variants in the code can be introduced, driving evolution

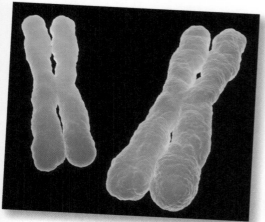

DIRECT ACTION

• • • • • • •

It was long thought that the wet, warm environment of a biological cell would prevent any quantum action by permanently imposing decoherence (see page 110). If this mechanism is experimentally verified, it's a powerful example of a quantum phenomenon with a direct effect on a macro object.

— and it seems that one potential mechanism for this mutation process is quantum tunneling.

The tread part, known as a "base pair," is linked by the electromagnetic attraction of hydrogen bonding. Each half of the tread ends in a proton, which is attracted to an atom — nitrogen or oxygen, for example — in the other half of the pair. But the distance involved is small and the proton, as a quantum particle, can tunnel from one half to the other, changing the chemical makeup of the DNA and causing a mutation.

◀ *Human chromosome 1 — DNA is usually wrapped on spindles, making it visible as a bundle under a microscope.*

Chromosomes are single DNA molecules. Human chromosome 1 (by no means the largest in nature) contains around 10 billion atoms.

▲ DNA's iconic helix form holds data in the "base pairs" linking the spirals.

▼ The proton can tunnel across hydrogen bonds, shown in the two types of base pairs as dotted lines.

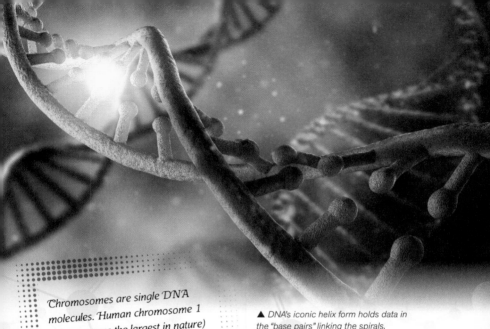

A·T base pair

G·C base pair

PHOTOSYNTHESIS

Of all the biological processes that are directly dependent on quantum effects, perhaps the most obvious are those that involve emitting or absorbing light.

The most familiar and essential of these is photosynthesis, the mechanism used by plants to turn light energy into chemical energy to keep the plant

Part of the photosynthesis process involves the fastest known chemical reaction, taking only a trillionth of a second.

▼ The green swath of a forest canopy is a natural power station thanks to the quantum effects in photosynthesis.

alive and growing. When a photon of light pushes up the energy of an electron in a chlorophyll molecule, we see the first quantum effect in a complex chain of processes.

To be used, the energy has to be transferred to a part of the plant cell called the photosynthetic reaction center. This is made possible by enabling the energy of the electron to be passed along, molecule to molecule, in a wavelike process. These waves are in step, making this process the biological equivalent of a laser, which produces light where the waves (or the phases, if considering light as photons) are similarly in step. The energy-passing process also seems to make use of the probabilistic nature of quantum events, testing the possibilities as if it were a quantum computer (see page 284) to find the best route to the processing center.

The flow of electrons powers a movement of protons across a membrane, driving the production of nature's energy storage molecule, ATP. When the electrons finally return to the chlorophyll, a side effect is the production of a molecule of oxygen as a waste product, resulting in the oxygen-rich atmosphere we need to breathe.

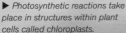

▶ *Photosynthetic reactions take place in structures within plant cells called chloroplasts.*

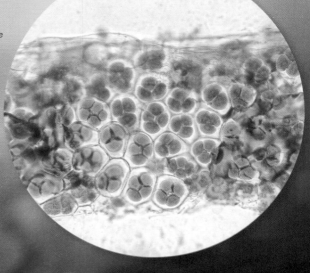

GUIDING THE PIGEON

A number of birds, such as the homing pigeon and the European robin, have a mysterious ability to navigate through the air even when they can't see any landmarks. For some time it has been known that this is due to their ability to pick up the Earth's magnetic field — the birds have a built-in compass — but the exact mechanism that would explain how this works has proved elusive.

However the process occurs, it is a quantum action. One of the better-supported suggestions is that the energy of incoming light is used to split a molecule into two charged parts called "free radicals." These are reactive compounds that, when uncontrolled, can cause cancer by damaging DNA, but here the spare electron on each of the pairs of free radicals acts like a tiny compass, because one of the quantum properties of an electron is its spin, and spin direction is influenced by magnetic fields (see page 152).

By having an impact on the interaction between the electron spin and that of the nucleus in chemicals in the eye called cryptochromes, the location of the bird within the Earth's magnetic field can change the way this chemical reacts, providing a pointer that could provide feedback, multiplied over numerous molecules, to the bird's brain.

FREE RADICALS

· · · · · · · ·

They may sound like a political faction, but free radicals are chemical compounds containing an atom, such as oxygen, that has unpaired electrons, which make them very good at reacting with other molecules. When free radicals occur in biological systems where they are not expected, they can damage other molecules by reacting with them. So, for instance, unwanted free radicals can damage the structure of DNA. However, they are used safely in many biological processes.

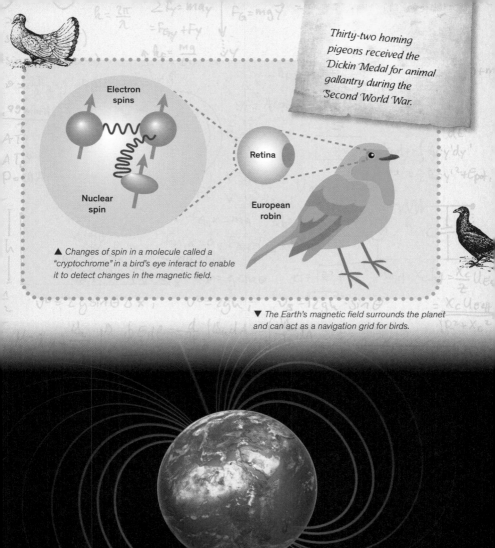

Thirty-two homing pigeons received the Dickin Medal for animal gallantry during the Second World War.

Electron spins

Retina

Nuclear spin

European robin

▲ Changes of spin in a molecule called a "cryptochrome" in a bird's eye interact to enable it to detect changes in the magnetic field.

▼ The Earth's magnetic field surrounds the planet and can act as a navigation grid for birds.

CHAPTER 9

LIVING THE QUANTUM

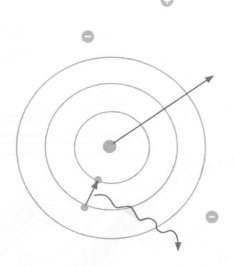

▶ We are surrounded by quantum technology, never more so than in the ubiquitous smartphone, which contains a wide range of quantum devices.

THE UBIQUITOUS QUANTUM

Quantum physics is fascinating in its own right — but it is also, alongside material science (itself increasingly driven by quantum discoveries), one of the most widely applied fields of physics, responsible by some estimates for around 35 percent of gross domestic product in technologically advanced countries.

▲ *Quantum devices in the home.*

As always, quantum physics underlies the nature of all matter and light, but here we see the benefit of making active use of quantum effects.

There has been a three-stage development of the way that practical scientists and engineers have addressed the quantum. Initially, quantum effects were used simply because they were part of nature. Anything done using materials, for instance, involved the manipulation of atoms. By the end of the 19th century, quantum effects were increasingly harnessed more directly, with electrons, for example, manipulated using magnetic and electrical fields. And since the 1950s, there have been conscious attempts to construct materials and products that harness quantum properties, in order to deliver remarkable new devices, from electronics to materials that are specially manufactured to have unusual quantum effects.

A modern smartphone contains a host of quantum technologies, from the screen and the flash memory to the GPS technology and the processor.

▼ *Every separate, functional part of a smartphone uses one or more quantum technologies.*

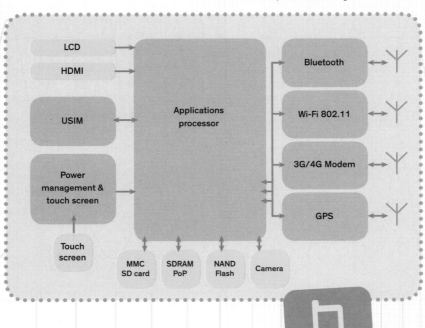

ACCIDENTAL QUANTUM

Although we have not been conscious of it, we have benefited from quantum effects for as long as we have existed. Our main source of energy, directly for heat and light, and indirectly as the power source of our foodstuffs and the weather, has always been the Sun — the biggest quantum-powered object within four light years of Earth.

MISSING QUANTA

It is arguably easier to say what *doesn't* make use, directly or indirectly, of quantum physics than what does. Apart from intangibles such as love and our appreciation of art (and even then, quantum processes in the brain are involved), the only significant exclusion is gravity. As we will discover (see page 296), even this may be brought within the quantum physics family, but as yet it has eluded all attempts to do so.

More mundanely, every chemical reaction we utilize, whether biological or external, such as producing a flame to cook or warm ourselves, involves quantum processes. As we have seen, chemical reactions are driven by atomic structure, itself a quantum concept. We have also made increasing use of magnetism and electricity over the centuries, unaware until very recently of the quantum nature of these phenomena.

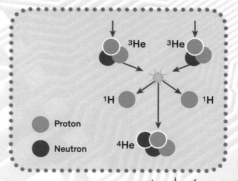

▶ *Without quantum tunneling, the positively charged nuclei in the Sun could not get close enough to fuse.*

Many scientists think that the biological processes creating consciousness involve quantum effects such as tunneling.

▶ Whether chemical reactions, magnetism or the plasma of a flame, quantum processes are at the heart of reality.

CROOKES TUBES

The second stage of our use of the quantum, based on electrons, can be seen clearly in the historically important Crookes tube, named for one of its key inventors, the British chemist and physicist William Crookes.

These were glass tubes that were sealed after most of the air in them had been removed. At one end of the tube would be a source of electrons, which were beamed toward the other end by putting a positive electrode partway along the tube, attracting the negative electrons.

The stream of electrons was known as a "cathode ray" because it emerged from the negatively charged electrode,

PAINTING WITH ELECTRONS

Soon, experimenters were using electrical and magnetic fields to shift the path of cathode rays. This was the mechanism behind the CRT (cathode-ray tube) TV, commonly used until the 1990s.

which was called a "cathode." Many electrons were absorbed by the positive electrode, but others shot past it to hit the glass. This glowed as the incoming particles boosted electrons in the atoms of the glass to a higher energy level, from which they then fell back, giving off light. Later, fluorescent material was painted onto the end of the tube to magnify the effect.

Initially, exactly what was behind the effects of the Crookes tube was uncertain, but with the discovery of the electron

▲ *The positively charged anode traditionally took the form of a Maltese cross.*

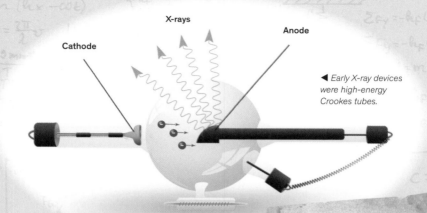

Cathode

X-rays

Anode

◀ Early X-ray devices were high-energy Crookes tubes.

(see page 40), it became clear that the process involved the manipulation of electrons, or electronics. In the tube we have both electromagnetic effects, in the acceleration of the electrons, and quantum electrodynamics, in the production of light when the electrons interact with the atoms in the glass.

By putting a very high voltage on the positive electrode, electrons could be accelerated fast enough that the glow produced was not visible light but X-rays. Early X-ray devices were simply variants of the Crookes tube.

▶ Until LCDs became standard, TVs and computer monitors were based on a sophisticated version of the Crookes tube.

▲ *Light bulbs needed to heat the filament until it was white hot.*

ELECTRONICS

We tend to assume that electronics refers to familiar solid-state devices, but it began with Crookes tubes. What was initially thought of as a ray, like a beam of light, was soon identified as a stream of electrons that could be manipulated using magnets and electrical fields. Electronics involves no more than the manipulation of electrons — when you switch on your TV, computer or phone, everything that happens internally is due to this manipulation.

In early Crookes tubes, the electrons were produced by an electrical discharge across the thin air in the tube, which ionized air molecules and dislodged electrons that were then attracted by the positive electrode. However, later tubes had a better vacuum — with less air in the tubes to deflect passing particles, more

electrons would make it to the other end. They used a more controlled method of production, with a filament like that in a traditional light bulb, heated until it released electrons into the vacuum.

The key to making the tube something more than a pretty effect was the manipulation of electromagnetic fields, especially by using electrically charged plates that could be switched on and off easily. The negatively charged stream of electrons, accelerated toward the positively charged anode, could then be switched, directed and controlled.

A traditional magnet could be used to give a fixed change of path to a beam of electrons, but by using a variable electrical charge or electromagnets, the path of the electrons could be steered from place to place. This made it possible to "write" on a screen with electrons, generating an image made of phosphorescent material that would glow long enough to sustain the image until the electrons were brought back to boost it once more.

Early light-bulb filaments were made of carbon, because metal filaments burned out too easily. But the filaments in a Crookes tube did not need to be heated to the same degree, so were made of metal.

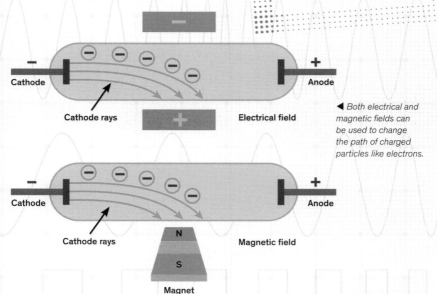

Cathode

Cathode rays

Electrical field

Anode

Cathode

Cathode rays

N

S

Magnet

Magnetic field

Anode

◀ Both electrical and magnetic fields can be used to change the path of charged particles like electrons.

THE VALVE

The key to making use of electronics is providing a switch or amplifying a signal — two of the most common features of an electronic circuit.

It would take only a small modification of the Crookes tube to produce a valve (known as a "vacuum tube" in the United States). In fact, early Crookes tubes were a crude form of the type of valve known as a "diode," through which electricity can flow in only one direction. Unlike a piece of wire, because of the charged electrodes, electrons could only flow from cathode to anode.

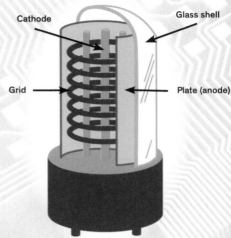

▲ *The triode valve was a compact tube with a third electrode, the grid, to control electron flow.*

To make them more compact, most valves were small glass cylinders with heaters or cathodes running up the middle. The grid (if present) was a tube with holes around it and the positive anode was a bigger tube around the outside.

However, the real breakthrough device was the triode. This involved putting another electrode between the negatively charged cathode and heater that produced the electrons, and the positively charged anode. This new electrode, with holes in it, was called a "grid." When the grid was not charged, a strong current flowed through the tube. But as the negative charge on the grid increased, fewer electrons flowed.

FLEXIBLE TRIODES

• • • • • • • • •

The triode could act as a switch,
turning the flow through the tube on
and off, or as an amplifier — a small
signal varying the strength of the
electrical field on the grid would be
duplicated as a stronger signal in the
flow of electrons.

◀ Array of valves in an ENIAC
computer from the 1940s.

SEMICONDUCTORS

Valves worked well, but they had a number of problems. Devices with a lot of valves, such as the early computers, gave off vast quantities of heat, which tended to stop the system working. Furthermore, valves were impossible to miniaturize below a certain point, and they were fragile and couldn't be switched on immediately as the heater needed time to warm up.

This meant that by the 1950s, experiments were under way to find materials that could perform the same switching and amplifying roles without the fragility, size and waste energy of valves. The obvious contenders were semiconductors. These materials,

◀ Insulators, such as the ceramic insulator on a power line, prevent the flow of electrons.

▼ Conductor, such as copper, allow a free flow of electrons.

◀ Semiconductors, such as silicon, allow a degree of electrical current to flow in ways that can be modified. Silicon is the most common semiconductor in electronic devices, deploying thin wafers of the crystalline material.

somewhere between a conductor such as a metal and an insulator such as glass, allow controlled flows of electrons — the essential requirement for reproducing the function of a valve.

Unlike the crude valves, designing semi-conductor components required explicit awareness of quantum physics to exploit the operation of the "band gap" — the jump made by an electron from being in orbit around an atom to being free to conduct.

Despite its name, the early cat's whisker radio did not involve cats — the "whisker" was a thin metal wire touching a semiconductor.

THE TRANSISTOR

The solid-state version of the triode was called the "transistor." It was developed at Bell Labs (then the research center of the American Telephone and Telegraph Company) in the late 1940s and came into practical use in the 1950s.

▲ The first transistor, which won its inventors — John Bardeen, William Shockley and Walter Brattain — the 1956 Nobel Prize in Physics.

Early transistors were made up of a three-way sandwich of semiconductors, with each part acting as the equivalent of one element of a triode valve. These sandwiches alternated different types of doping.

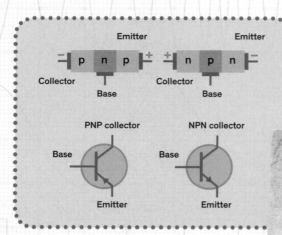

◀ The most basic transistors consist of simple junctions between n and p doped materials.

The first patent for a transistor was filed as early as 1926, but it was not practical to make one with the materials of the day.

Doping can use an element that either adds extra electrons (called "n-type" because it enhances the negative) or has fewer electrons available than the native semiconductor ("p-type," for positive). For example, with silicon, the n-type doping agent might be phosphorus and the p-type boron.

Just like the grid in the triode valve, the central slice of a semiconductor transistor (n in a pnp version or p in an npn version) controls the flow of electrical current between the other two slices, providing switching or amplifying capabilities.

THE IMPORTANCE OF HOLES

It might seem there is little value in p-type doping, but the missing electrons make it easier for electrons to flow from elsewhere. The missing electrons are referred to as "holes," often treated as if they were actual objects, as dealing with a relatively small number of holes rather than a far larger number of electrons can make calculations easier.

▶ The first domestic commercial application was the transistor radio, which made radios portable. They became so commonplace that they were simply called "transistors."

◀ Although modern transistors are mostly found in integrated circuits (see page 258), there are still applications for individual transistor devices.

ALL IN ONE

The earliest transistors were much smaller than valves, which typically ranged from ⁴⁄₅ inch (2 cm) to 4 inches (10 cm) in height, but were still around ²⁄₅ inch (1 cm) across. Constructing a modern PC processor, with its 1.4 billion transistors, would consume impossibly large quantities of time and space.

By the late 1950s, just as transistors were becoming widely used, monolithic circuits that included multiple components such as transistors started to become practical. Rather than build individual components, the integrated circuit starts with a wafer of silicon on which a layer of silicon dioxide is grown. By then spraying on an extremely thin metal or polycrystalline silicon layer, it is possible to produce the same effect as a transistor but in a far more compact form.

▲ Large wafers of silicon are grown to act as the base for multiple integrated circuits.

There are two benefits from this chip-based approach. First, after the initial design, the circuit can be constructed much faster, meaning electronics could be taken to the mass market. And second, far more components can be crammed in at little extra cost. In a modern integrated circuit, quantum technology is employed in multiple ways to reproduce the large individual components as tiny segments.

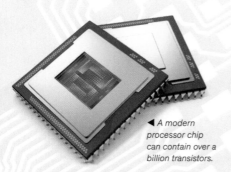

◀ A modern processor chip can contain over a billion transistors.

MOORE'S LAW

• • • • • •

The effectiveness of the approach
is demonstrated by Moore's law, an
observation by Intel founder Gordon
Moore that the number of transistors
in a chip doubles approximately
every two years. This has held true
for over four decades.

It's not clear who
invented the integrated
circuit, but the honor is
usually shared between
Jack Kilby of Texas
Instruments and Robert
Noyce of Fairchild
Semiconductor.

INVENTING THE LASER

The laser is the archetypal quantum device. Inside is a "lasing" material, in which a photon of light pushes up the energy of an electron in an atom. A second photon then hits the electron and — rather than being absorbed — triggers the release of the first photon.

The lasing material acts as a light amplifier, getting one photon in and sending two out. And the photons produced are in phase, which means that light is emitted in a tight beam.

ANOTHER EINSTEIN INSPIRATION

In 1916 Albert Einstein developed the basic theory behind the laser, known as "stimulated emission of radiation." Hence the name: light amplification by stimulated emission of radiation.

▲ *Theodore Maiman invented the laser in May 1960.*

Basic laser technology uses a lasing material, such as a ruby, in a chamber between a pair of mirrors. Light is flashed into the chamber and runs back and forth between the mirrors, building up in-phase photons. However, the second mirror is only partially silvered, which allows some of the photons out to produce the laser beam.

Three American scientists led the race to construct the laser. Charles Townes had an early theoretical lead, but was sidetracked by an overly complex mechanism as he decided — incorrectly — that ruby lasers were impractical and developed a device using a gas as lasing material. Gordon Gould had the first practical design, but his progress was hampered by security problems as his project was sponsored by the U.S. Defense Department's Advanced Research Projects Agency. So it was Theodore Maiman who succeeded in producing the first working laser in May 1960.

Maiman's laser owed a lot to a development in photography. One of his coworkers had recently brought in one of the new electronic flash units that were beginning to replace flashbulbs, and this proved the ideal light source to pump light into the ruby lasing material.

When Gordon Gould was working on his laser, his company received funding from the U.S. Department of Defense. This meant that Gould had to be security cleared. He failed, partly because two of his referees had beards and were therefore considered subversive. As a result, he was banned from even reading his own notebooks.

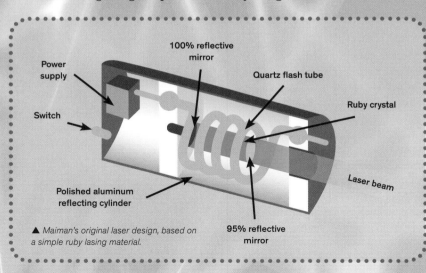

▲ Maiman's original laser design, based on a simple ruby lasing material.

Power supply

Switch

100% reflective mirror

Quartz flash tube

Ruby crystal

Laser beam

Polished aluminum reflecting cylinder

95% reflective mirror

▲ *Fiber optics make beautiful light displays, but more importantly provide the backbone of our modern communication systems.*

LASERS IN USE

Of the three key players in the development of lasers, Townes and Maiman both focused on communications technology, where the devices have proved essential in the development of the high-speed internet and long-distance communications, both of which use laser light in fiber optics.

Gould had made a more dramatic sales pitch, predicting military applications such as laser-guided weapons and the use of lasers themselves as weapons, something that is only now being

THE SKIPPING INTERNET

In 1870, English physicist John Tyndall demonstrated the "fountain of light," in which a stream of water channeled light that was fed into the holding tank. Similarly, in fiber optics, the laser light in a glass fiber repeatedly hits the side and is reflected, skipping along the fiber's curves.

developed, particularly for ships, where the absence of recoil in a laser weapon is ideal.

No one, however, foresaw the full breadth of laser use. Whether it is in medical applications such as laser surgery, reading optical discs such as CDs, DVDs and Blu-rays, or in printers and bar code scanners, lasers have become a part of everyday life in a way that was never expected.

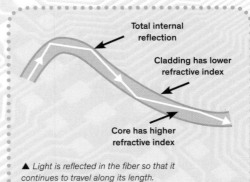

Total internal reflection

Cladding has lower refractive index

Core has higher refractive index

▲ *Light is reflected in the fiber so that it continues to travel along its length.*

"Gould was the first to use the term "laser." Townes, the earliest of the three scientists to start work, always called it an "optical maser," as his big success had been with the microwave equivalent.

◀ *Unlike CDs or laser printers, the laser beam in a supermarket scanner is visible to the user.*

HOLOGRAMS

The hologram is one early application of lasers that is still to reach its full potential. Looking at a photograph of a landscape is clearly different from looking at the real view through a similarly sized piece of glass. Though the idea of looking at the countryside through a small piece of glass may seem strange, the process involved is important: the photons from the objects we see through the glass arrive from different directions with different phases. As we move our viewing position, we get a changing viewpoint and a

Hungarian-British scientist Dennis Gábor patented the hologram in 1947. A working model could not be built until 1964, following the construction of the first laser.

different image. A hologram captures this information about the photons arriving at a surface, such as a sheet of glass, and reproduces that information, projecting it from the surface, resulting in a three-dimensional image.

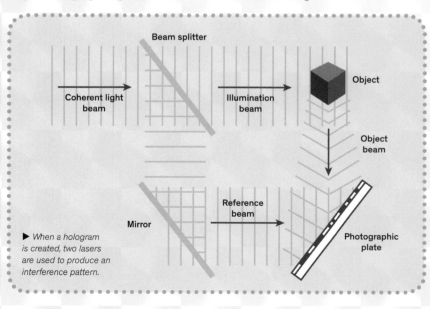

Beam splitter

Coherent light beam

Illumination beam

Object

Object beam

Mirror

Reference beam

Photographic plate

▶ When a hologram is created, two lasers are used to produce an interference pattern.

The effect is captured using two lasers (or a single source that is split), one to illuminate the view and the other shining directly on the recording surface. The holographic process captures the interference patterns between the two lasers, recreating the view when a third laser is shone on the patterns.

Although there have been significant advances in hologram technology, the capture process remains intrusive as it requires scanning with lasers rather than using natural light, and it is slow, so holograms continue to have no more than a niche role.

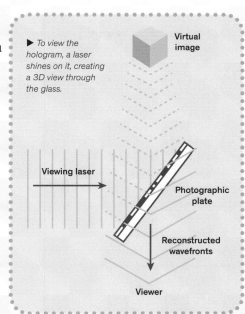

▶ To view the hologram, a laser shines on it, creating a 3D view through the glass.

Virtual image

Viewing laser

Photographic plate

Reconstructed wavefronts

Viewer

PRINTED HOLOGRAMS

The holograms in credit cards and banknotes work by reflecting light off foil and playing back the interference pattern between two or more transparent layers above the foil. The result is less complete than a traditional transmission hologram.

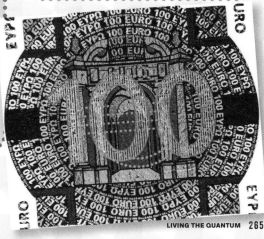

▶ The most familiar holograms are those used for security markers.

THE ELECTRON MICROSCOPE

When Gábor first came up with the idea of holography he was thinking in terms of another quantum breakthrough — the electron microscope — which it was hoped would provide a wider range of imaging. Indeed, the electron microscope has proved a valuable application of quantum physics since the 1930s.

With the idea that electrons could act as waves came the notion of replacing photons in a microscope with electrons. This had a big advantage, because a traditional microscope is limited in terms of how small an object it can view by the wavelength of the light. It is impossible to resolve an image of less than around a wavelength across. As electrons can have wavelengths at least 1,000 times smaller than those of visible light, this makes possible a significant improvement in imaging.

ELECTRON VARIANTS

The first electron microscopes were modeled on the traditional optical microscope, passing a beam through a semitransparent object, but over time a range of electron microscopes were developed, providing different types of imaging. A further type of quantum microscope, the scanning tunneling microscope, which uses a quantum interaction between a probe and the surface of the object being viewed, can produce images down to the level of individual atoms. Another impressive capability of the scanning tunneling microscope is the interaction between the probe and the surface to move individual atoms on a sample. This was impressively demonstrated in 1989 when researchers spelled out the initials IBM in 35 xenon atoms on a nickel crystal.

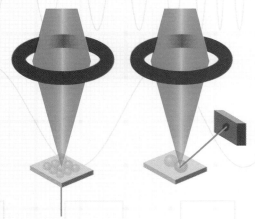

◀ Whereas in the transmission electron microscope the electron beam passes through the object, the scanning electron microscope produces secondary electrons (or photons) when it hits the surface, which are picked up by surrounding detectors.

◀ The electron microscope scans its sample using a beam of electrons.

While light microscopes are limited to magnifications of around 2,000, in principle an electron microscope can reach magnification of 10 million.

▶ Objects viewed through an electron microscope reveal an alien landscape of submicroscopic structures.

MRI SCANNERS

One of the most quantum-loaded pieces of machinery, now in everyday use in hospitals, is the magnetic resonance imaging (MRI) scanner, developed in the 1970s. Originally known as "nuclear magnetic resonance" (NMR) scanners, the name was changed because of the misleading negative connotations of "nuclear."

The mechanism for these scanners is a quantum effect. The human body contains lots of water, each molecule of which has a pair of separate protons. When a patient enters the scanner, an extremely powerful magnet changes the quantum spin of these protons. When the magnetic field is switched off, the protons revert back to their original spin state, emitting photons at radio frequencies as they do so. The affected water molecules become miniature radio transmitters and their output is picked up by receivers placed around the patient.

Generating this effect requires very powerful magnets, which in turn require cooling down to extremely low temperatures, usually with liquid helium, to make them superconductors (see page 220). This means that both the mechanism of the scan, using quantum spin, and the technology that flips the spin, using superconductivity, are direct applications of quantum effects.

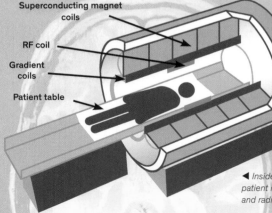

Superconducting magnet coils

RF coil

Gradient coils

Patient table

◄ *Inside the MRI scanner, the patient is surrounded by magnets and radio receivers.*

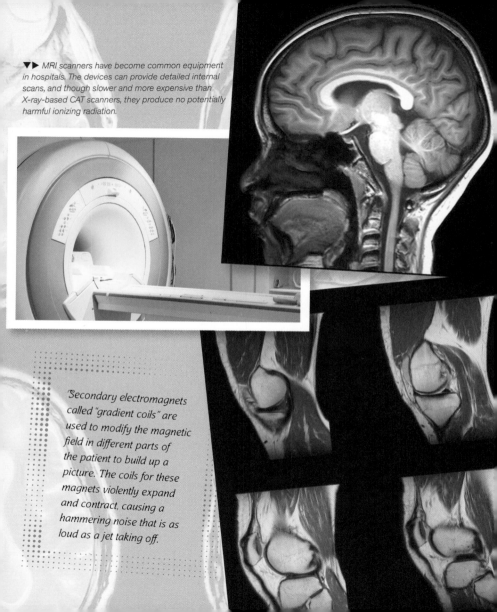

▼▶ MRI scanners have become common equipment in hospitals. The devices can provide detailed internal scans, and though slower and more expensive than X-ray-based CAT scanners, they produce no potentially harmful ionizing radiation.

Secondary electromagnets called "gradient coils" are used to modify the magnetic field in different parts of the patient to build up a picture. The coils for these magnets violently expand and contract, causing a hammering noise that is as loud as a jet taking off.

LEVITATING TRAINS

Although superconductors are yet to find a wide commercial application, one exciting possibility is to use superconducting magnets to support extremely high-speed trains.

Conventional trains are limited by friction and stresses between the wheels of the train and the tracks. However, a maglev (magnetic levitation) train uses the repulsive force of a magnet to float the train above the track, slashing its resistance to motion.

Getting hundreds of tons of train to float takes a lot of doing — only superconducting magnets can produce sufficient lift. By sequencing changes in the magnetic field, the same magnets can also be used to push the train along the track, as if it were a particle passing through an accelerator.

MAKING MAGLEV REAL

Although scientists have been conducting maglev experiments for decades, there have been very few commercial ventures to date. Safety concerns and the costs of the current generation of superconducting magnets have held back widespread use of the technology, which has been limited to small-scale prototypes. However, the first major commercial line, the Chuo Shinkansen in Japan, is now under construction, with completion scheduled for the 2030s.

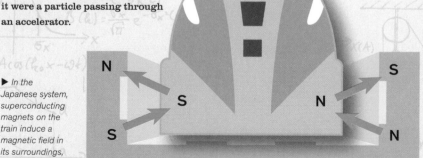

► In the Japanese system, superconducting magnets on the train induce a magnetic field in its surroundings, causing levitation.

◀ The first commercial maglev train ran between Birmingham Airport and Birmingham International Station from 1984 to 1995.

Conventional high-speed rail operates at around 155 miles per hour (250 km/h), while prototype Japanese maglev trains have reached over 370 miles per hour (600 km/h).

▼ The Chuo Shinkansen maglev line will initially run between Tokyo and Nagoya and eventually on to Osaka.

FLASH MEMORY

The strange phenomenon of quantum tunneling (see page 114) is not just responsible for the functioning of the Sun. It is at the heart of a piece of technology that most of us have in the house: flash memory.

Computer memory relies on the value of an electrical charge to register as a zero or a one, and a constant power supply is usually required to keep that memory stable. However, flash memory — the memory found in mobile phones, solid-state computer disks and memory sticks — retains its information even when the power is switched off.

Computer memory comes in units of one bit, which is a store that can have the value of either zero or one. Each of these bits of memory is represented by the state of a transistor operating as a switch — it has one value if a current is flowing and another if it cannot flow. Flash memory uses a special type of transistor called a floating gate transistor. Here the "gate" that acts as the switch is isolated in the middle of an insulator. Whether or not the gate is charged provides the switch function, but it's not possible to charge or discharge the gate directly. Instead, quantum tunneling is used to switch its state, crossing the barrier of the insulation.

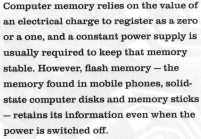

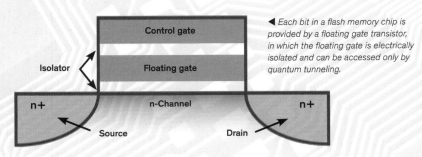

◀ Each bit in a flash memory chip is provided by a floating gate transistor, in which the floating gate is electrically isolated and can be accessed only by quantum tunneling.

Control gate

Floating gate

Isolator

n+   **n-Channel**   **n+**

Source   **Drain**

Flash memory was initially relatively slow and expensive, making it limited to specialist applications. However, it has now become a far better substitute for magnetic hard disk drives as it is both quicker to access and much more robust, so better suited for portable equipment.

Flash memory was invented in the early 1980s by Japanese electronics engineer Fujio Masuoka while working at Toshiba.

► *All removable memory cards contain flash memory.*

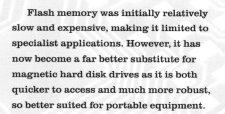

▼ *In a solid-state drive (SSD) the fragile rotating magnetic disk of a hard disk is replaced by flash memory chips.*

QUANTUM PHOTOGRAPHY

There are few industries that have been so totally transformed by quantum physics as photography. You have only to look at the trajectory of Kodak from world-spanning business to protective bankruptcy to see how the rise of the digital camera has changed our lives.

A quantum technology lay behind Kodak's problems. There are two approaches to digital photography: CMOS (complementary metal oxide semiconductor) and CCD (charge-coupled device). The cheaper and more common CMOS contains an array of light-sensitive circuits able to detect light levels (the light is passed through tiny red/green/blue filters to separate the different colors for each pixel). By contrast, the CCD, which also uses color filters, has an array of tiny electron collectors, each notching up the number of electrons displaced by incoming photons. In effect, in a CCD each pixel has an "electron bucket" in which the electrons it receives are counted up to produce the final picture.

CHOOSING YOUR SENSOR

CCD was the earlier of the technologies, but it has been replaced for most cameras by the cheaper CMOS. However, CCD has some technical advantages (CMOS devices can often collect information from only one row of sensors at a time, making them poor at capturing fast-moving images) and remains the professional technology of choice.

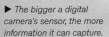

▶ *The bigger a digital camera's sensor, the more information it can capture.*

▲ While CCD was traditionally used in digital SLR cameras because of better quality images, CMOS has become more common as the technology has improved.

In 1975, Kodak was the first company to develop a digital camera, but it suppressed the technology, aware of what it would do to the business.

▼ Comparison of the features of CCD and CMOS.

Description	CCD	CMOS
Camera components	Sensor + optic support chips + optics	Sensor + sometimes optic support chips
Speed	Medium to fast	Fast
Sensitivity	High	Low
Noise	Low	Medium
System complexity	High	Low
Sensor complexity	Low	High
Fill factor	High	Low
Chip output	Voltage (analog)	Bits (digital)
Pixel signal	Electron	Voltage
Uniform shuttering	Medium to high	Low

JOSEPHSON JUNCTIONS AND SQUIDS

Another wide-ranging, if less familiar, quantum technology is the Josephson junction, named after the Cambridge physicist Brian Josephson. The junction, which is microscopic in size, consists of a pair of superconductors with a barrier between them. When quantum tunneling takes place across this barrier it has unique properties — notably, the junction acts as an extremely sensitive voltage detector when an AC (alternating current) supply is applied.

Josephson junctions have a number of sensing applications, but they most frequently appear in the form of SQUIDs — superconducting quantum interference devices — which use Josephson junctions to detect tiny changes in magnetic fields.

Like lasers, SQUIDs are finding more and more applications. For example, they make excellent detectors in MRI scanners and magnetic-field microscopes, and they are now used in specialist detectors that can pick up

◀ Brian Josephson shared the 1973 Nobel Prize in Physics for "his theoretical predictions of the properties of a supercurrent through a tunnel barrier, in particular those phenomena which are generally known as the Josephson effects".

Brian Josephson is often at odds with the scientific community because of his support for fringe ideas such as telepathy and the memory of water.

small changes in the Earth's magnetic field.
These variations indicate the presence
of metallic objects, such as unexploded
bombs, and, unlike their conventional
equivalents, SQUIDs can detect through
water as well as through earth.

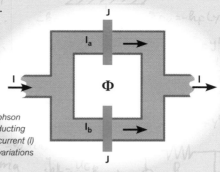

▶ In a simple SQUID, a pair of Josephson
junctions (J) between two superconducting
bars (dark grey) change the level of current (I)
passing through in response to tiny variations
in the magnetic field detected (Φ).

▼ Demonstration of one of the
first SQUID magnetometers.

QUANTUM OPTICS

We are used to optics that manipulate beams of light through the use of mirrors and lenses, whether in spectacles, telescopes or microscopes. As we have seen, these are quantum devices. However, it is also possible to control the flow of light using equipment that makes use of other quantum effects. Some quantum optics can reproduce the effects of electronics using photons, so the field is known as "photonics."

This wide field includes devices that produce light by explicit quantum means. (Of course, a traditional light bulb *does* uses a quantum effect, but knowledge of quantum physics was not required to design one.) The laser is one such device (see page 260), while another, the light-emitting diode (LED), is now used in most screens and as a low-energy replacement for conventional bulbs. In the LED, as electrons from one part of the diode drop into holes in the other part, they release energy in the form of photons of light.

Other specialist quantum optics use totally new materials, such as metamaterials

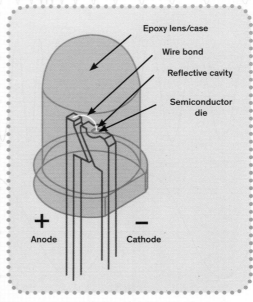

Epoxy lens/case

Wire bond

Reflective cavity

Semiconductor die

+
Anode

—
Cathode

▲ *The LED bulb has transformed lighting, wasting far less energy (in the form of heat) than conventional light bulbs.*

▲ LED bulbs are rapidly replacing compact fluorescents as they use less electricity, come to full power immediately and are much less fragile.

— substances that have properties not found in nature. For example, transparent materials usually have a positive refractive index, bending light inward as it enters. But some metamaterials have a negative refractive index, bending light the opposite way. This makes it possible to construct super-lenses with incredibly high magnification and cloaking devices that bend light around an object.

Harry Potter doesn't need to worry. Although metamaterials can cloak an object, they don't yet work with visible light, and the object needs to be as small as the light's wavelength. The first effective cloak worked only with microwaves, working on an object no more than a few centimeters across. Even so, the principle allows for genuine invisibility.

QUANTUM DOTS

Some quantum technology hides coyly behind innocent-sounding terms such as "electronics," but other forms proudly announce their quantum heritage.

One of the difficulties of working with a quantum particle is simply keeping hold of it. This is particularly important if individual quantum particles are to be used in large-scale devices. A quantum dot is a tiny solid-state semiconductor device that can be used as a single-electron transistor or to generate extremely small light sources, in effect operating as if the dots were artificial atoms that give off photons in response to stimulation.

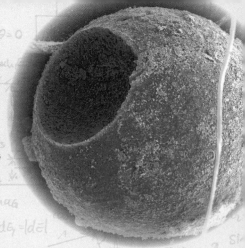

▲ Some quantum dots are grown as tiny crystals. This one is made from cadmium sulfide.

Single-electron transistors are miniature versions of flash memory (see page 272), where an electrical charge is held in an isolated store accessed by quantum tunneling. In this case, that store (called the "island") is a quantum dot and holds only a single electron. The

▶ In a single-electron transistor, the quantum dot (QD) interfaces the gate to the main electrodes.

Drain

Gate | Quantum dot

Tunnel junctions

Gate capacitor

Source

quantum dot represents the final stage of miniaturization of electronics, because once a system is working with single electrons at a time, there is nowhere smaller to go. If devices are to become any smaller, they will have to switch to an alternative, such as photonics, where photons replace electrons.

The quantum dot is not only one of the smallest conceivable electronic components, but its ability to hold a single electron means that it could be used to construct a special kind of quantum device known as a "qubit" (see page 282).

▼ When subject to an electrical current, quantum dot crystals emit light of a color that is dependent on their size.

QUBITS

"Qubit" is a mash-up of "quantum" and "bit." In our computers and phones, bits are the smallest level of storage, each holding a value of 0 or 1, but in a qubit, the bit is a quantum particle — and this changes everything.

The reason for this is superposition (see page 152). A quantum property such as spin exists as a combination of probabilities — for example, 60 percent spin up, 40 percent spin down. This superposition means that a qubit has a flexible range of values — it can represent any length of decimal number. And as qubits are combined, the values that can be dealt with increase exponentially. With just a few hundred working qubits, it would be possible to build a quantum computer (see page 284) capable of performing calculations beyond any current computer.

That's the good news. The downside is that qubits are tricky. If, for example,

STORING REAL NUMBERS

Imagine the two percentages for the superposition of spin up and spin down as a direction, from 100 percent up to 100 percent down. Each direction can be assigned a value between 0 (fully up) and 1 (fully down). The superposition represents a true decimal value.

we checked the value of the qubit in the superposition above, we would get "up" 60 percent of the time, and "down" 40 percent of the time, so we would get only a single value each time.

A wide range of options are being considered for qubits. The most popular are probably photons, electrons and quantum dots, but nuclear spin used by MRI machines and a special variant of the Josephson junction are among other contenders.

According to American physicist Benjamin Schumacher, who came up with the term, "qubit" was a play on the old forearm-based measurement, the cubit.

▲ Although qubits themselves are invisibly small quantum particles, the environment to support them is currently imposing: this device houses just five qubits.

▶ Because the qubit stores the superposition of two values, it can represent any direction between up and down, which can represent any decimal fraction.

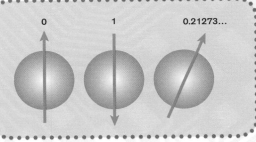

0   1   0.21273...

QUANTUM COMPUTERS

Manipulating a collection of bits makes it possible to build conventional computers. Unsurprisingly, qubits are the working components of quantum computers.

This technology has already been under development for decades, and though it is proving extremely difficult to make a commercially viable quantum computer, hundreds of teams are working on it.

The difficulties of getting quantum computers to work range from decoherence (see page 110), because it's hard to keep the qubits from interacting with their environment, to simply getting information to pass around them, for which quantum teleportation is essential (see page 174). However, a working quantum computer requires far fewer qubits than a conventional computer has bits, and significant progress is being made. It seems likely that quantum computers will have a big impact in the next few years.

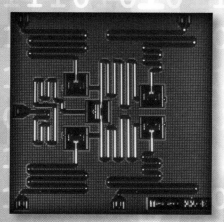

▲ *A close-up of the circuit design and layout for IBM's five-qubit superconducting quantum computing device.*

At least two programs written to run on quantum computers date back to the 1990s. One can find the factors of large numbers, meaning it has the ability to break current computer encryption, while the other can make searches that take the square root of the time the process would take on a conventional computer.

▼ The D-Wave processor looks similar to a conventional integrated circuit, but contains around 512 of its alternative-style qubits.

RIDING THE D-WAVE

Although true quantum computers are not yet commercially available, one device, the D-Wave, claims to be a working quantum computer. It certainly uses quantum effects, which can perform some calculations far faster than a conventional computer, but it employs a highly specialized process called "quantum annealing" that means it will always be limited in application.

▼ Crystal core of a quantum computer, at high magnification.

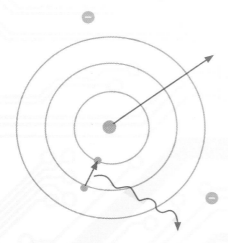

CHAPTER 10
QUANTUM UNIVERSE

▶ The Hubble Telescope's detection of quantum particles — photons — only scratches the surface of the involvement of quantum physics on the universal scale.

COSMIC QUANTUM THEORY

Quantum physics is usually described as the science of the very small — electrons, atoms, photons — all far smaller than the objects in our everyday lives. However, this does not mean that we can forget the impact of the quantum as we look to the universe and take in its cosmological implications.

It seems strange that quantum particles behave so differently from familiar objects, even though those objects are made up entirely of quantum particles. Yet, as we have seen, we can't escape quantum effects, even on the macro scale.

When we look beyond the Earth to the wider universe, and start to consider very large phenomena, we can't put quantum physics to one side. Although gravity is the most significant force on the astronomical scale, the other three fundamental forces have important roles to play, whether in the nuclear reactions of a star or the electromagnetic pulses that blast out from solar flares. Quantum effects also crop up in a number of hypothetical concepts, such as black holes and the big bang. Indeed, it's hard to find anything in which quantum theory does not play a part.

THE MISSING FORCE
• • • • • • • • • •

Gravity doesn't fit with the quantum world — for the time being. In Einstein's masterwork, the field equations for gravity that lie at the heart of his general theory of relativity are not quantized. Although most physicists believe wholeheartedly that there ought to be some way to combine gravity with the other forces of nature, which are quantum-based, as yet this has proved elusive.

Einstein spent most of the last 30 years of his life looking for a way to pull together gravity and the other forces of nature, but failed.

THE FOUR FORCES

1. Electromagnetic (quantized)

2. Strong nuclear (quantized)

3. Weak nuclear (quantized)

4. Gravity

NO THEORY OF EVERYTHING

Physicists like to point out that we have a GUT but no TOE. That's a "grand unified theory" pulling together the three quantized forces — electromagnetism plus the strong and weak nuclear forces — but not a "theory of everything" folding gravity into the mix.

Electromagnetism and the weak force were first combined in the "electroweak" theory, before the strong force was added. Current theory suggests that in the early universe these forces were combined into one, but as the universe cooled, the different forces split off in a process known as "symmetry breaking."

Einstein's general relativity works very effectively to describe the behavior of large bodies, while quantum theory dominates the very small. Although there is no reason why gravity has to fit in with such a unified theory, physicists like the simplicity (of the concept if not the math) that this implies.

▶ Symmetry breaking occurs when a very small change results in a random change of state. For example, a pencil balanced on its tip could fall symmetrically in any direction, but once it moves slightly the symmetry is broken and it changes to a new state — fallen over.

SOMETHING NEW

• • • • • • • •

As it stands, quantum theory and general relativity are incompatible. To achieve a unified theory would require modifying one or both of them — yet each works strikingly well. It's entirely possible that a unified theory would have to come from a whole new direction, replacing both of these existing theories.

Three physicists — the Americans Sheldon Glashow and Steven Weinberg and the Pakistani Abdus Salam — shared the 1979 Nobel Prize in Physics for the electroweak theory.

▶ (left to right) Glashow, Weinberg and Salam.

BIG BANG

Since the early 1930s, when it became clear that the universe was expanding, it has seemed reasonable to believe that by working back in time we would come to the point when it all started.

The term "big bang" is used loosely to describe this model, though purists point out that it doesn't describe the start of the universe, but rather its sudden expansion.

LOCATING THE BIG BANG

. . . .

We tend to think of the big bang as happening at a specific point at the center of the universe. But in fact it occurred at the end of your nose. The expansion of the universe is the expansion of space itself, not an expansion in space. So *every* point in the current universe is where the big bang occurred.

▼ *Much of the early life of the universe, shown here evolving from the big bang on the left, was dominated by quantum effects.*

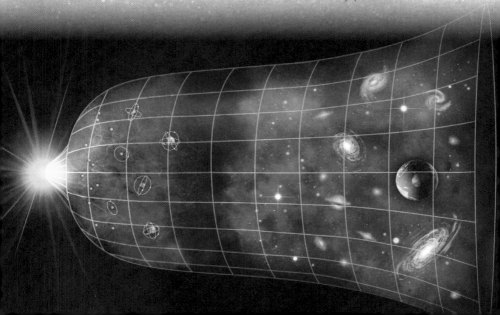

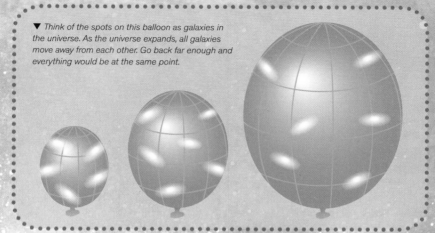

▼ *Think of the spots on this balloon as galaxies in the universe. As the universe expands, all galaxies move away from each other. Go back far enough and everything would be at the same point.*

The most extreme version of big bang theory has the universe starting as a "singularity" — a point of infinite density from which everything emerged. In practice, most agree that the appearance of infinity here suggests that current theory doesn't apply, but even a finite and very small start implies a starting condition of the universe that was subject to quantum physics.

The big bang theory works backward in time from the current expanding universe to get to a starting point where it could be no smaller. Latest estimates put that point around 13.8 billion years in the past. The original big bang theory had problems because the universe appears to be too uniform. To make the cosmologists' assumptions work, it was suggested that there was a sudden and drastic expansion of the universe when it was only a tiny fraction of a second old, during which the universe increased in volume by a factor of over a trillion trillion. Such a process, known as cosmic inflation, would also explain how small quantum fluctuations in the early universe became the seeds of the large-scale structures we see today.

English physicist Fred Hoyle coined the term "big bang" for a theory he didn't support in a radio broadcast in 1948.

BLACK HOLES

Stars naturally collapse under the pull of gravity — it's how they formed from a cloud of gas in the first place. But as they collapse they heat up and the heat energy, which causes the atoms to jostle around, eventually balances out the collapse. One of the first deductions from the general theory of relativity was that as stars run out of energy, some will start to collapse again, getting smaller and smaller until they disappear to a point.

Such a "disappearance to a point" would result in a singularity. Like the big bang, this suggests a situation that current theory cannot explain, so something else must apply. However, there is good astronomical evidence that bodies behaving in a very similar fashion to black holes *do* exist.

Because general relativity shows that matter produces a warp in space–time that results in the effects of gravity, get close enough to a black hole and the warping is so severe that nothing, not even light, can escape. The sphere within which this occurs is known as the "event horizon."

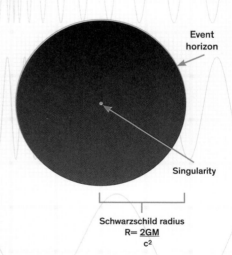

Event horizon

Singularity

Schwarzschild radius
$$R = \frac{2GM}{c^2}$$

▲ *The Schwarzschild radius, named after Karl Schwarzschild, the German physicist who came up with a solution to Einstein's equations, defines the size of the event horizon for a particular mass of matter.*

▲ There appears to be a very massive object, identified as a black hole, in the center of our Milky Way galaxy.

VISIBLE BLACKNESS

• • • • • • • • • •

Although nothing gets out of black holes, they still give off light. Any nearby material that is pulled into them will accelerate, emitting light as it does so. And there will be a glow from a process known as Hawking radiation, in which pairs of quantum particles that pop into existence as a result of the uncertainty principle (see page 100) are split by the event horizon.

English astronomer John Michell, born in 1724, came up with a concept similar to that of the black hole. He was working on escape velocity, the speed at which something thrown from the surface of a body would need to start to escape the gravitational pull. This speed increases as the mass of the body increases. Michell reasoned that an extremely massive star would have an escape velocity that was greater than the speed of light.

QUANTUM GRAVITY

If gravity is ever to be brought into a theory of everything, it will likely have to be quantized — coming in chunks, rather than continuously variable, just as, for instance, electromagnetism is quantized into photons.

The field equations of general relativity currently have no degree of uncertainty, none of the probabilistic element that is now so familiar in quantum theory. This could not be the case if we were to establish a quantum theory of gravity. Similarly, quantizing gravity would mean that we would expect there to be a force-carrier particle or particles, as there are for the other forces, and it is

likely that space and time would also be quantized — meaning that they are not continuous, but granular, with a minimum size below which nothing can be determined.

If space–time were quantized, a quantum (the smallest quantity) of space may have the dimensions of the Planck length (named after Max Planck — see page 68), which is around 6×10^{-34} inches (1.6×10^{-35} m). This is something like 10^{25} times *smaller* than the hydrogen atom, the smallest atom. And the time equivalent is around 5.4×10^{-44} seconds. These Planck values were originally conceived as a way to have units that were derived from fundamental constants of nature. For example, the Planck length is the square root of Planck's constant (see page 70) multiplied by Newton's gravitational constant, divided by the cube of the speed of light.

Once both Einstein's general theory of relativity and quantum electrodynamics

If the Planck length is a reasonable guide to size of the quanta of space, there is finally a use for the number the googol (10^{100}), because there would be around a googol quanta in a cubic meter of space.

became well established, it seemed an obvious next step to combine the two and bring quantum theory into the fold. Einstein spent decades attempting to do so, as did many of his colleagues. The obvious starting point was to treat gravity using the same kind of quantum field theory that had been so effective for electromagnetism. But this approach proved to have much bigger problems with renormalization — the emergence of unwanted infinities — than had QED. If there were some sort of quantum gravity particle, there seemed no way to avoid these interacting with each other and producing a runaway gravitational collapse.

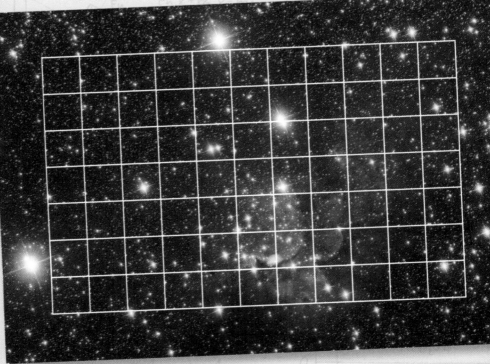

▼ *If space–time is quantized, space is no longer continuous but consists of tiny grains.*

GRAVITATIONAL WAVES

Black holes weren't the only prediction of the general theory of relativity that could be tested. Einstein also showed that sudden movements of massive objects should produce a wave in space–time, as an oscillating object does in a pool of water. This is a particularly interesting phenomenon for quantum gravity when you consider the relationship between waves and particles in quantum physics.

There have been many attempts over time to detect gravity waves, but it's not easy for a number of reasons. While light is mostly relatively high frequency (visible light ranges from 4×10^{14} to 8×10^{14} hertz), gravity waves are expected to

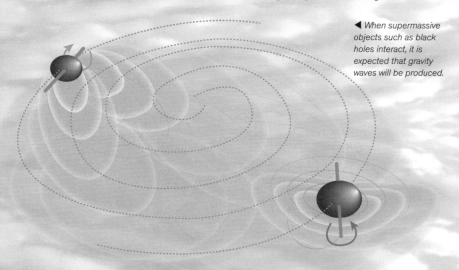

◀ When supermassive objects such as black holes interact, it is expected that gravity waves will be produced.

have a frequency of between 0.001 and 10,000 hertz, making their detection problematic, because they could easily be confused with earthbound vibration. Gravity-wave detectors tend to look for variations in the length of a measuring device — originally metal bars and later beams of light. The existence of gravity waves proved elusive for decades, but in 2015, the LIGO (Laser Interferometer Gravitational-Wave Observatory) detector picked up a signal that in 2016 was confirmed to be caused by the first directly detected gravity waves.

▲ The Livingston LIGO, showing the 2½-mile-long (4 km) interferometer arms.

LIGO uses a pair of devices 1,864 miles (3,000 km) apart, one in Hanford, Washington, and the other in Livingston, Louisiana, comparing their outputs to produce its results.

◀ The first generation of astronomical telescopes improved on the naked eye by magnifying optical observations.

A NEW ASTRONOMY

When the LIGO discovery was announced, the press made much of it "confirming Einstein's general relativity." While it's true that the general theory of relativity predicted the existence of gravity waves, the theory itself had already been tested extensively, and no confirmation was required. Instead, the discovery of gravity waves was significant because it may create a new medium for astronomy.

For a very long time, astronomy used visible light. We looked at the sky, initially with the naked eye and then using lenses and mirrors. From the 20th century onward, other parts of the electromagnetic spectrum have been brought into play, using a range of instruments, from radio telescopes to X-ray and gamma-ray observatories. However, all these telescopes have their limitations.

As we look farther out into space, we look farther back in time, because light takes time to reach us. (This is also true of gravity waves, which travel at the speed of light.) And if we look far enough, we hit a barrier. Until the universe was around 370,000 years old, it was opaque. Light could not pass through it. But gravity waves could — and this is why they may provide the basis for a whole new type of astronomy. The 2015 LIGO observation, for example, gave us our first-ever direct observation of black holes.

▲ Second-generation telescopes added other parts of the electromagnetic spectrum, such as radio, to our observations.

▶ Third-generation telescopes moved into space to avoid the distortion of the Earth's atmosphere. Gravity waves now give us a whole new way to observe the universe.

THE GRAVITON

Although a full quantum theory of gravity has proved elusive, it is possible to treat gravity waves as a stream of particles. This approach only deals with gravity waves, much as it was possible to treat light waves as particles before there was a complete theory of quantum electrodynamics.

▼ *If gravitons could be made compatible with the general theory of relativity, we would expect there to be a stream of them between orbiting bodies such as the Earth and the Moon.*

The Russian physicists Dmitri Blokhintsev and F.M. Gal'perin gave the hypothetical gravitational particle the name "graviton" in 1934. However, it wasn't until the late 1950s

▼ As gravity travels at the speed of light, if the Sun were to vanish, it
would take these lengths of time before the planets ceased to feel its
pull. The figures in the middle column below represent the distance
from the Sun in astronomical units (AU).

Mercury	0.387	193.0 seconds or 3.2 minutes
Venus	0.723	360.0 seconds or 6.0 minutes
Earth	1.000	499.0 seconds or 8.3 minutes
Mars	1.523	759.9 seconds or 12.7 minutes
Jupiter	5.203	2,595.0 seconds or 43.3 minutes
Saturn	9.538	4,759.0 seconds or 79.3 minutes
Uranus	19.819	9,575.0 seconds or 159.6 minutes
Neptune	30.058	14,998.0 seconds or 4.2 hours

*Gravity travels at
the speed of light, so
if the sun suddenly
vanished, we would still
feel its effect for around
eight minutes as the
gravity continued to
head our way.*

that Peter Bergmann in the United
States and Paul Dirac in the United
Kingdom worked out the mathematics
behind the particle.

Just as the photon is both the
quantized version of a light wave and
the carrier of electromagnetic force,
the graviton would be the quantized
version of a gravity wave and the carrier
of gravitational force — should that ever
prove possible to quantize. Also like a
photon, the graviton would be massless
(otherwise gravitational force would
be limited in range, and there would be
interesting issues of self-action as the
graviton would somehow have to have

REAL OR UNREAL?

• • • • • • • • •

Gravitons remain hypothetical.
Although those working on quantum
gravity assumed that, like quantum
field theory, problems of unwanted
infinities in the calculations could
be resolved, in practice this has
proved impossible. Therefore, a new
theory is required.

its own mini gravitons to provide its
gravity, which themselves would need
their own... and so on).

STRING THEORY

String theory was first developed in an attempt to explain the standard model of particle theory. Extended to take in gravitational effects, it has become the most widely researched attempt to link quantum physics and gravity.

When simply described, string theory seems beguilingly attractive. There is no longer a zoo of different quantum particles (see page 196). Instead, each particle is just a different mode of vibration of the one true fundamental particle — the string. This is a one-dimensional construct that can be either open or formed into a loop that vibrates in different ways.

Unfortunately, although the basic concept is easy to grasp, the mathematics is anything but approachable, and string theory works only if there are nine spatial dimensions. If that weren't enough of a problem, there's the matter of deriving anything useful from the theory. String theory's equations don't have a single solution — they have around 10^{500} solutions. This is such a problem that some leading physicists argue that string theory is not truly science at all.

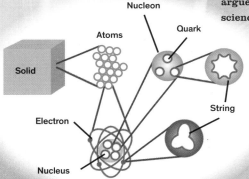

◀ *In string theory there is an extra level to the hierarchy of matter below the quark and the electron.*

German physicist Martin Bojowald has said that string theory is certainly a theory of everything, because everything — and anything — can happen within it.

▲ *There are far more solutions to string theory than stars in the observable universe.*

Open string

Closed string

▲ *In string theory, strings can be closed, with no ends, or open, with two ends. Each can vibrate in different ways.*

THE TESTABILITY REQUIREMENT

Unless a scientific theory makes predictions that can be tested, it has no value. But string theory has so many possible solutions that testing it is incredibly difficult.

▶ *The top row shows some of the open string vibrations, and the bottom some of the closed vibration modes, each corresponding to a particle.*

MANY DIMENSIONS

Mathematicians and physicists are used to working in multiple dimensions.

In mathematics you can have as many dimensions as you like, and in quantum physics it is not unusual to consider an artificial multidimensional space where each dimension is the equivalent of one possible outcome, or of a different variable in an equation. But these are just mathematical tools. String theory is different — it requires that there are six extra *real* spatial dimensions.

The obvious question is: where are these dimensions? Even if we struggle to

▲ *Just as we can have a two-dimensional projection of a cube, the tesseract is a three-dimensional projection of a four-dimensional hypercube. It has eight "sides," each an identically sized cube, but perspective drawing distorts them.*

IMAGINING DIMENSIONS

Think of the dimensions of a square drawn on paper. Now add a third dimension at right angles to the first two. That represents a three-dimensional object. To get your head around string theory, you need to carry on adding extra dimensions at right angles until you have nine of them.

envisage what a nine-dimensional world would be like, we would expect to see some impact from the extra dimensions on everyday life — but we don't. Envisaging a fourth spatial dimension is not too difficult. In fact, it would make it possible to have a finite universe with no boundaries. Think of the surface of the Earth as a two-dimensional surface. By bending it through a third dimension, the finite surface of the Earth has no boundaries. Similarly, if the universe were folded through a fourth dimension, we needn't have the complexity of an edge.

However, getting to the full dimensional scale of string theory needs many more dimensions to be tucked away. To get around this, the unseen dimensions are assumed to be curled up in such tight loops that we cannot detect them.

British physicist Paul Davies has said: "Maybe string theorists... one day may be able to tell us how it works. Or maybe they are away in Never-Never Land."

▼ We are used to three dimensions of space and one of time. String theory needs a far more complex canvas.

M-THEORY

String theory has five major variants, and these have been combined into a single approach known as M-theory. This super-theory was introduced by American physicist Edward Witten in 1995. However, there is a price to pay for this amalgamation — yet another spatial dimension is required. So M-theory needs a total of 10 dimensions of space plus one of time.

In M-theory the basic unit, rather than a string, is a brane, which can have any number of dimensions up to 10. The one-dimensional form of this, twisted through other spatial dimensions, simplifies to a string.

The extra dimensions in M-theory (or string theory, for that matter) are quite different from the science-fiction concept of a "parallel dimension," which is more like an alternative universe alongside

LIVING ON A BRANE

In M-theory our universe can be represented by a membrane — or "brane" — floating in the 10-dimensional space. An alternative to the conventional big bang theory, called "ekpyrotic theory," suggests that the universe began expanding when two of these branes collided.

▶ Taken to the extreme, a rolled-up two-dimensional sheet appears one-dimensional.

our own. But the extra dimensions in M-theory are just the same kind of dimensions as the ones we currently experience, which means they have to be modified in some way to explain their apparent absence.

Like its string theory constituents, M-theory requires the extra spatial dimensions to be curled up. Imagine, for example, a two-dimensional sheet of paper. Curl it up very tightly until it forms a long, thin tube. Seen from a distance, it will appear to be one-dimensional. Now imagine a 10-dimensional piece of paper with seven of those dimensions rolled up and you would see our familiar three-dimensional universe.

▶ "Ekpyrotic" means "out of the fire." In the M-theory-based ekpyrotic universe, multiple big bangs can occur as the brane we occupy collides with others in 10-dimensional space.

No one is sure what the "M" stands for. Witten has suggested each of "magic," "membrane" and "mystery."

LOOP QUANTUM GRAVITY

The greatest challenge to string theory is loop quantum gravity. This doesn't require extra dimensions, and doesn't see the string as a super-particle. Instead, loop quantum gravity breaks down space–time itself into a kind of atom — the loop in the theory's name.

Once quantum theory is applied to space–time, the uncertainty principle comes into play (see page 100). Rather than momentum and position being linked here, an equivalent pairing would

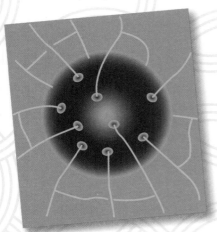

▲ Loop quantum gravity offers a possible solution to entropy problems with a black hole's event horizon, represented in this artist's impression as quantum loops puncturing the horizon.

SPACE–TIME

• • • • •

Einstein's earliest venture into relativity, the special theory of relativity, introduced the concept of space–time. Special relativity shows that movement through space inevitably influences time, making it impossible to think of space and time as separate. Instead, time is added as a kind of fourth dimension to make space–time.

be an area of space and its curvature. It's common to use a sheet of rubber as an analogy for space–time in general relativity, as general relativity requires space and time to be warped, rather as a sheet of rubber distorts when a heavy object is placed on it. The equivalent analogy for loop quantum gravity is a weave of loops.

When envisioning the loops in this theory it's hard to avoid thinking of a structure within space and time. But the weave of loops *is* space–time (or, more precisely, space, because time is an added extra in this theory). Empty space would still have loops, while a region with no loops would contain no space.

Electron neutrino   Electron antineutrino

Positron   Electron

Down quark   Up quark

▲ In one version of loop quantum gravity, particles are represented by twists in the loops of space, where different "braids" appear as familiar particles.

It's entirely possible that neither this approach nor string theory will survive — but physicists will continue trying to bring gravity into the quantum family that underpins everything else in the universe.

There seem to be three possible outcomes. A theory such as M-theory or loop quantum gravity could bring general relativity into the quantum arena. Or one of the general theories of relativity and quantum theory will be replaced with something new. Or, finally, it may prove impossible ever to perform such a unification. There is no reason why there should be a single approach covering everything, but all our experience so far suggests that we will reach a more universal view.

Whatever the final outcome, quantum theory has proved superbly accurate in describing the behavior of matter and light, and it will continue to enable us to produce remarkable new technologies for centuries to come.

Loops, just like quantum particles, are not clearly defined. Rather, they are fuzzy clouds of probability.

INDEX

PICTURE CREDITS

CHAPTER 4

p.91 © Science Source/ Science Photo Library

p.92 © American Institute of Physics/Science Photo Library

p.93 © David Parker/ Science Photo Library

pp.94–5, 100–1, 112–13 © Inga Nielsen/ Dreamstime.com

pp.95, 101, 102, 105, 108, 115, 117 illustrations by Geoff Borin

p.97 © Science Photo Library

p.99 © Gift of Jost Lemmerich/Emilio Segre Visual Archives/ American Institute of Physics/Science Photo Library

pp.99, 100 © Emilio Segre Visual Archives/ American Institute of Physics/Science Photo Library

p.103 Wikimedia/Koogid

pp.105, 107 Wikimedia

p.109 © Keystone/Stringer/ Getty Images

p.116 with kind permission of Prof Rolf Pelster, University of Saarbrücken

CHAPTER 5

p.120 © CERN

pp.121, 123, 125, 127, 128, 136, 143, 144, 147, 148 illustrations by Geoff Borin

p.121, 126 Wikimedia

p.122 Wikipedia

p.124 © Lucas Taylor/CERN

p.130 (L) © Harvey of Pasadena/American Institute of Physics/ Science Photo Library

p.130 (M) © Science Photo Library

p.130 (R) © Emilio Segre Visual Archives/American Institute of Physics/ Science Photo Library

p.131 © Diana Walker/Getty Images

p.132 © Estate of Francis Bello/Science Photo Library

p.142 Natural History Museum, London/Science Photo Library

pp.148–9 © Inga Nielsen/ Dreamstime.com

CHAPTER 6

pp.153, 155, 156, 160, 165, 166, 172 illustrations by Geoff Borin

p.158 Photograph by Paul Ehrenfest, copyright status unknown. Colored by Science Photo Library

pp.158–9 © Inga Nielsen/ Dreamstime.com

p.161 (L) © Science Source/ Science Photo Library

p.161 (M, R) © Emilio Segre Visual Archives/ American Institute of Physics/Science Photo Library

p.165 © Peter Menzel/ Science Photo Library

p.167 Thaler Tamás/ Wikimedia Commons

p.173 © Jin Liwang/ Xinhua/eyevine

CHAPTER 7

pp.181, 183, 186–7, 188, 193, 197, 199, 204, 206 illustrations by Geoff Borin

p.179 © Christophe Vander Eeecken/Reporters/ Science Photo Library

p.180 © NASA/WMAP Science Team/Science Photo Library

pp.180–1, 202–3, 208–9 © Inga Nielsen/ Dreamstime.com

CHAPTER 10

pp.290–1, 296–7,
308–9 © Inga Nielsen/
Dreamstime.com
p.291 (L) © Emilio Segre
Visual Archives/
American Institute of
Physics/Science Photo
Library
p.291 (M,R) © CERN/
Science Photo Library
pp.293, 294, 298, 304,
305, 308, 310, 311
illustrations by Geoff
Borin
p.299 © LIGO
p.306 © Equinox Graphics/
Science Photo Library